Cloud Native Automation with Google Cloud Build

Easily automate tasks in a fully managed, scalable, and secure platform

Anthony Bushong

Kent Hua

BIRMINGHAM—MUMBAI

Cloud Native Automation with Google Cloud Build

Associate Group Product Manager: Rahul Nair
Publishing Product Manager: Niranjan Naikwadi
Senior Editor: Athikho Rishana
Content Development Editor: Divya Vijayan
Technical Editor: Rajat Sharma
Copy Editor: Safis Editing
Project Coordinator: Ashwin Kharwa
Proofreader: Safis Editing
Indexer: Hemangini Bari
Production Designer: Prashant Ghare
Marketing Coordinators: Nimisha Dua

First published: October 2022
Production reference: 1160922

Published by Packt Publishing Ltd.
Livery Place
35 Livery Street
Birmingham
B3 2PB, UK.

ISBN 978-1-80181-670-0

www.packt.com

For my good amid these, Margaux and Benjamin, and the verse you help me contribute.

- Anthony

To my wife, Lien-ting, and children Mason and Madison, thank you for your support and patience to help make this a reality.

- Kent

Contributors

About the authors

Anthony Bushong is a senior developer relations engineer at Google. Formerly a field engineer and practice lead for Kubernetes and GKE, he worked with companies implementing automation for their Kubernetes infrastructure in Cloud Build – since version 1.3! He now focuses on distilling those experiences into multiple formats of technical content – all with the goal of teaching and enabling people, no matter their background or experience.

I want to thank my mom, dad, grandma, family, and friends – all without whom I am nothing. I also want to thank those who have taken a chance on me in this industry, recognizing that much of what I have been able to achieve has been built on your trust and teachings – especially Diane Anderson and Andrew Milo.

Kent Hua is a global solutions manager focused on application modernization. He has years of experience with customers modernizing enterprise applications focusing on both business and technical challenges on-premises and the public cloud. Over the years, he has helped organizations decompose monoliths and implement microservice patterns wherever applicable into containers running on Kubernetes. While enabling these organizations, he has identified culture and the automation of processes as critical elements in their modernization journeys.

I want to thank my parents, family, friends, and colleagues who have made me who I am today. Through our interactions and experiences, not a day passes without learning something new.

About the reviewers

Marcelo Costa is a technology lover, who over the past ten years has spent quite some time working across both software and data roles. Between working in companies and personal projects, he has worked with multiple technologies and business areas, always looking for challenging problems to solve.

He is a cloud computing **Google Developer Expert** (**GDE**) and likes to help the community by sharing knowledge with others through articles, tutorials, and open source code. He currently works as a founding engineer at Alvin, an Estonian startup in the data space.

Damith Karunaratne is a group product manager at Google Cloud, who focuses on driving continuous integration and software supply chain security efforts. Prior to joining Google, Damith spent over a decade helping companies of all shapes and sizes solve complex software challenges, including leading various CI/CD and DevOps initiatives.

Damith earned a bachelor's degree in science with a computer science emphasis from McMaster University.

Table of Contents

Part 3: Practical Applications

7

8

9

10

Part 4: Looking Forward

11

Preface

This book gets started by discussing the value of managed services and how they can help organizations focus on the business problems at hand. Build pipelines are critical in organizations because they help build, test, and validate code before it is deployed into production environments.

We then jump right into Cloud Build: the fundamentals, configuration options, compute options, build execution, build triggering, and build security.

The book's remaining chapters close with practical examples of how to use Cloud Build in automation scenarios. While Cloud Build can help with software build life cycles, it can also coordinate delivery to runtimes such as serverless and Kubernetes.

Who this book is for

This book is for cloud engineers and DevOps engineers who manage cloud environments and desire to automate workflows in a fully managed, scalable, and secure platform. It is assumed that you have an understanding of cloud fundamentals, software delivery, and containerization fundamentals.

What this book covers

Chapter 1, Introducing Google Cloud Build. Establish the foundation of serverless and managed services, focusing software build life cycles with Cloud Build.

Chapter 2, Configuring Cloud Build Workers. It's a managed service, but we still need compute and this chapter discusses the compute options available.

Chapter 3, Getting Started – Which Build Information Is Available to Me?. Kicking off the first build and discovering the information available once it has started to help inform you of success or debug issues.

Chapter 4, Build Configuration and Schema. You can get started quickly with Cloud Build, but knowing the configuration options can help you save time.

Chapter 5, Triggering Builds. This is the critical component for automation: react when something happens to your source files or trigger from existing automation tools.

Chapter 6, Managing Environment Security. It's a managed service, but there are still shared responsibilities, determining who can execute pipelines, what pipelines have access to, and how to securely integrate with other services.

Chapter 7, Automating Deployment with Terraform and Cloud Build. We can also leverage Cloud Build to automate Terraform manifests to build out infrastructure.

Chapter 8, Securing Software Delivery to GKE with Cloud Build. Discovering patterns and capabilities available in Google Cloud for secure container delivery to Google Kubernetes Engine.

Chapter 9, Automating Serverless with Cloud Build. It's serverless, but we still need to automate getting from source code to something running.

Chapter 10, Running Operations for Cloud Build in Production. Additional considerations may need to be made when preparing for Cloud Build in production, working with multiple teams.

Chapter 11, Looking Forward in Cloud Build. What's next? For instance, we'll look at how Cloud Deploy leverages Cloud Build under the hood.

To get the most out of this book

You need to have experience with software development, software delivery, and pipelines to get the most out of the book and Cloud Build.

You will need access to a Google Cloud account. Examples in the book can be performed in Google Cloud's Cloud Shell, which has the majority of the tools and binaries noted in the book. If external resources are needed, they are noted in the respective chapters.

If you are using the digital version of this book, we advise you to type the code yourself or access the code from the book's GitHub repository (a link is available in the next section). Doing so will help you avoid any potential errors related to the copying and pasting of code.

Download the example code files

The book will leverage examples from different repositories noted in the respective chapters.

Code examples in the book can be found at `https://github.com/PacktPublishing/Cloud-Native-Automation-With-Google-Cloud-Build`. If there's an update to the code, it will be updated in the GitHub repository.

Packt Publishing also have other code bundles from our rich catalog of books and videos available at `https://github.com/PacktPublishing/`. Check them out!

Download the color images

We also provide a PDF file that has color images of the screenshots and diagrams used in this book. You can download it here: `https://packt.link/C5G3h`.

Conventions used

There are a number of text conventions used throughout this book.

`Code in text`: Indicates code words in text, database table names, folder names, filenames, file extensions, pathnames, dummy URLs, user input, and Twitter handles. Here is an example: "In this case, we will be creating a private pool of workers that have the `e2-standard-2` machine type, with 100 GB of network-attached SSD, and located in `us-west1`."

A block of code is set as follows:

```
# Docker Build
- name: 'gcr.io/cloud-builders/docker'
  args: ['build', '-t',
         'us-central1-docker.pkg.dev/${PROJECT_ID}/image-repo/myimage',
         '.']
```

When we wish to draw your attention to a particular part of a code block, the relevant lines or items are set in bold:

```
. . .
INFO[0002] No cached layer found for cmd RUN npm install
INFO[0002] Unpacking rootfs as cmd COPY package*.json ./
requires it.
. . .
INFO[0019] Taking snapshot of files...
```

Any command-line input or output is written as follows:

```
$ project_id=$(gcloud config get-value project)
$ vpc_name=packt-cloudbuild-sandbox-vpc
```

Bold: Indicates a new term, an important word, or words that you see onscreen. For instance, words in menus or dialog boxes appear in **bold**. Here is an example: "Select **System info** from the **Administration** panel."

> Tips or important notes
> Appear like this.

Get in touch

Feedback from our readers is always welcome.

General feedback: If you have questions about any aspect of this book, email us at `customercare@packtpub.com` and mention the book title in the subject of your message.

Errata: Although we have taken every care to ensure the accuracy of our content, mistakes do happen. If you have found a mistake in this book, we would be grateful if you would report this to us. Please visit `www.packtpub.com/support/errata` and fill in the form.

Piracy: If you come across any illegal copies of our works in any form on the internet, we would be grateful if you would provide us with the location address or website name. Please contact us at `copyright@packt.com` with a link to the material.

If you are interested in becoming an author: If there is a topic that you have expertise in and you are interested in either writing or contributing to a book, please visit `authors.packtpub.com`.

Share Your Thoughts

Once you've read *Cloud Native Automation with Google Cloud Build*, we'd love to hear your thoughts! Scan the QR code below to go straight to the Amazon review page for this book and share your feedback.

`https://packt.link/r/1801816700`

Your review is important to us and the tech community and will help us make sure we're delivering excellent quality content.

Part 1: The Fundamentals

This part of the book will introduce you to Cloud Build. You will understand the context in which Cloud Build exists and brings users value, the core user journey when using Cloud Build, the architecture and environment in which builds run, and the means by which you can inspect and review a build execution.

This part comprises the following chapters:

- *Chapter 1, Introducing Google Cloud Build*
- *Chapter 2, Configuring Cloud Build Workers*
- *Chapter 3, Getting Started – Which Build Information Is Available to Me?*

Introducing Google Cloud Build

1

To properly introduce Google Cloud Build and the value it provides to its users, it's important to review the value that automation brings to IT organizations for common workflows such as cloud infrastructure provisioning and software delivery.

Automating these tasks may be helpful in increasing developer productivity for organizations; doing so with a managed service enables this productivity at a lower cost of operation, allowing individuals and teams to focus on the business task at hand, rather than managing all the infrastructure that runs the automation. There has been an increase in automation needs for processing AI/ML types of workloads (`https://cloud.google.com/architecture/mlops-continuous-delivery-and-automation-pipelines-in-machine-learning`), which are beyond the typical developer workflows. We will be focusing on the developer automation workflow (that is, continuous integration) aspects of Cloud Build in this book.

In this chapter, we will review Google Cloud Build through this lens, specifically discussing the following:

- The value of automation
- Before there was the cloud
- Reducing toil with managed services
- Cloud-native automation with Google Cloud Build

Technical requirements

- Data center and infrastructure concepts
- Public cloud concepts
- Software build concepts

The value of automation

The compilation of applications and services comes in all shapes and sizes. It may seem straightforward that code just becomes a packaged artifact, but for many scenarios, builds can have a number of complex steps with many dependencies. The steps involved in creating and testing an artifact may be manual, automated, or a combination of both to form a build pipeline. The following figure demonstrates an example build pipeline with a set of activities critical to the building of an application:

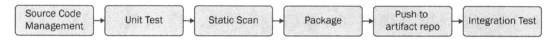

Figure 1.1 – Example build pipeline

Running these builds manually could potentially lead to careless mistakes with an outcome that may not be consistent or repeatable. When code is being built, it is very important to document even the smallest changes that were made to the build that resulted in something working as opposed to not working. This is where the use of a **source code management** (**SCM**) system, such as Git (`https://git-scm.com`), becomes critical in our overall pipeline. If a change was actually the result of a build step changed locally, not being able to repeat this can result in frustration and productivity loss.

This is especially relevant from the perspective of handing off work to a colleague. Having to understand and tweak a set of manual steps in a build would not be a good use of that colleague's time, when they could instead be focused on the code at hand. The time of each member of an organization is valuable and it's best to allow that individual to focus on being productive. This could be during a production outage, where time is best spent trying to fix the root cause of the outage rather than analyzing how to actually build the code. Depending on the impact of the outage, every second could have a monetary impact on the organization. In cases of a simple development handoff to a production outage, automation of a build would be very beneficial to the situation.

Imagine if a developer could solely focus on code development, rather than analyzing manual steps or difficult-to-execute builds. An organization might have automation in place, but it must be seamless for the developer in order to maximize productivity. This can be in the form of the following:

- Coding standards
- Boilerplate code
- Blueprints on how to use the pipeline

The preceding reference points could also apply to both what is considered the automation of the inner loop and the outer loop of software development. The inner loop of development typically consists of local development by a developer. Once code is completed in the inner loop, a merge request is created for addition to an integration branch. Once merged in an integration branch, this is where the typical build pipeline starts: the outer loop. Providing a starting point in the form of standards itself may not be automation; however, it could be baked into the configuration files. It may just be a starting point, a foundation that can also provide a level of flexibility for the developer to apply specific preferences.

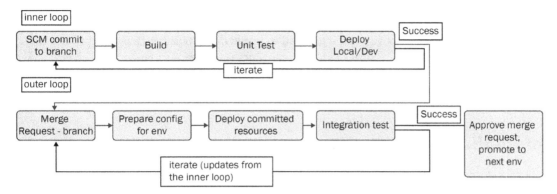

Figure 1.2 – Example inner and outer loops

The ecosystem of tools and integration that has been built around Git has helped drive the importance of version controlling not only source code but also configurations that define a build pipeline. GitOps (`https://www.weave.works/blog/the-history-of-gitops`) primarily focuses on **infrastructure as code** (**IaC**), ensuring that a runtime environment represents configurations declaratively stored in Git. The common use of Git tooling across developer and operation teams reduces the amount of friction for onboarding, which makes GitOps also critical for end-to-end automation.

Automation helped reduce end-to-end deployment times for this organization: `https://cloud.google.com/customers/reeport/`.

Once automation is streamlined, the team that owns the pipeline is able to aggregate metrics in order to determine areas for improvement at scale. This becomes a more achievable task, compared to when builds are executed manually by each developer. As mentioned earlier, pipelines in an organization could also include manual steps. These metrics could identify patterns where manual steps could possibly be automated. Reducing manual steps would increase the efficiency of a pipeline while also reducing the potential human errors that can occur. There may be situations where manual steps aren't automatable, but identification is key so that it can be considered in the future or to allow for teams to focus on other steps that can be improved.

This can reduce developers' frustration and improve overall productivity across teams, which can benefit the organization in the following ways:

- Delivering features faster
- Reducing the amount of time it takes to resolve issues

- Allowing teams to focus on other business-critical activities
- Feedback for continuous improvement of the pipeline

The value of automation can help an organization in many aspects. While metrics can be manually gathered, they can be most effective when aggregated in an automated pipeline. Decisions can be made to determine the most effective changes to a build pipeline. The metrics gathered from frameworks in place, such as GitOps, can also help feed into improving the end-to-end pipeline, not just the automation of source code compilation. Continuous improvement becomes more achievable when an organization can use metrics for data-driven decisions.

Before there was the cloud

There are a variety of tools on the market, ranging from open source to closed source and self-managed to hosted offerings, supporting build pipelines. Availability of the pipeline solution is critical in ensuring that code is built in a timely manner; otherwise, it may impact the productivity of multiple teams. Organizations may have separate teams that are responsible for maintaining the solution that executes the build pipeline.

Making sure there are enough resources

For self-managed solutions, the maintenance includes the underlying infrastructure, OS, tools, and libraries that make up the pipeline infrastructure. Scale is also a factor for build pipelines; depending on the complexity, organizations may have multiple concurrent builds occurring at the same time. Build pipelines need at least compute, memory, and disk to execute, typically referred to as workers within the build platform. A build pipeline may consist of multiple jobs, steps, or tasks to complete a particular pipeline to be executed. The workers are assigned tasks to complete from the build platform. Workers need to be made available so that they can be assigned tasks and such tasks are executed. Similar to capacity planning and sizing needs for applications, enough compute, memory, storage, or any other resource for workers must be planned out.

There must be enough hardware to handle builds at the peak. Peak is an important topic because in a data center scenario, hardware resources can be somewhat finite because it takes time to acquire and set up the hardware. Technologies such as virtualization have given us the ability to overprovision compute resources, but at some point, physical hardware becomes the bottleneck for growth if our build needs become more demanding. While an organization needs to size for peak, that also means that builds are not always running constantly at peak to make full use of the allocated resources. Virtualization, as mentioned previously, may help us with other workloads consuming compute during off-peak time, but this may require significant coordination efforts throughout the organization. We may be left with underutilized and wasted resources.

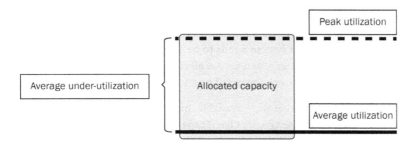

Figure 1.3 – Under-utilized resources when allocating for peak utilization

Who needs to manage all of this?

A team typically maintains and manages the build infrastructure within an organization. This team may be dedicated to ensuring the environment is available, resources are kept up to date, and new capabilities are added to support organizational needs. Requirements can come from all directions, such as developers, operators, platform administrators, and infrastructure administrators. Different build and pipeline tools on the market do help to facilitate some of this by offering plugins and extensions to extend capabilities. For instance, Jenkins has community contributions of 1,800+ plugins (https://plugins.jenkins.io/) at the time of writing this book. While that is quite an amount, this can also mean teams have to ensure plugins are updated and keep up to date with the plugins' life cycles. For instance, if the plugin is no longer being maintained, what are the alternatives? If multiple plugins perform similar functions, which one should be chosen? A rich community is beneficial as popular plugins bubble up and may have better support.

While productivity is impacted as mentioned, not having enough capacity or improperly sizing the build infrastructure could lead to slower builds. Builds come in all shapes; they can run in seconds for some, while to others, they can take hours. For builds that take hours or a long time, this would mean the developer and many other downstream teams are waiting. Just because a build is submitted successfully, it does not mean it completes successfully too; it could possibly fail at any point of the build, leading to lost time.

The team that is responsible for managing the build infrastructure may also be likely responsible for maintaining a **service-level agreement** (**SLA**) to the users of the system. The design of the solution may also have been designed by another team. As noted earlier, if builds are not running, there may be a cost associated because it impacts the productivity of developers, delays in product releases, or delays in pushing out critical patches to the system. This needs to be taken into account when self-managing a solution. While this was the norm for much of the industry before there was the cloud, in an on-premises enterprise, vendors developed tools and platforms to ease the burden of infrastructure management. **Managed service providers** (**MSPs**) also provided tooling layers to help organizations manage compute resources, but organizations still had to take into account resources that were being spun up or down.

Security is a critical factor to be considered when organizations need to manage their own software components on top of infrastructure or the entire stack. It's not just the vulnerability of the code itself being built, but the underlying build system needs to be securely maintained as well. In the last few years, a significant vulnerability was exposed across all industries (`https://orangematter.solarwinds.com/2021/05/07/an-investigative-update-of-the-cyberattack/`).

Eventually, when public cloud resources were available, much of the similar patterns discussed could be used – in this case, **infrastructure as a service (IaaS)** offerings in a cloud provider for handling the compute infrastructure. These eased the burden of having to deal with compute resources, but again, like MSPs, the notion of workers had to be determined and managed.

Organizations have had to deal with the build software pipeline platform regardless of whether the infrastructure was managed on-premises in their data center, in a co-location, or by an IaaS provider. It is critical to ensure that the platform is available and has sufficient capacity for workers to complete associated tasks. In many organizations, this consisted of dedicated teams that managed the infrastructure or teams that wore multiple hats to ensure the build platform was operational.

Reducing toil with managed services

In the previous section, we discussed the efforts involved in maintaining a platform for building applications and services. Many of the activities described in making sure the environment is always up and running could involve some toil. For example, Google's SRE handbook (`https://sre.google/sre-book/eliminating-toil/`) goes further into the elements of IT tasks that could be considered toil.

If we are able to avoid toil and know that a provider manages the underlying build infrastructure, we are able to focus on what is more important, the application that helps drive our business. This is one of the goals of managed services, letting the provider handle the underlying details, providing a consistent syntax that becomes the common language between teams, providing compute resources as needed, and not billing when the service is not being utilized.

It is one less component of a build pipeline to consider as the provider is maintaining the underlying infrastructure and they are able to provide the team with scale when needed at any given time. The MSP would be responsible for making sure that there are enough workers in order to execute all the jobs in the build pipeline. However, managed services could also be seen as a form of lock-in to a particular vendor or cloud provider. In most cases, a managed service typically has the best integration to services provided by the offering provider. This is where adding additional capabilities are much more streamlined, but not limited, to the following:

- Triggering mechanisms
- Secrets management

- Securing communication and data transfer between integrated services
- Observability

The integrations are there to help save time and, in reference to the original theme of this book, allow an organization to focus on the application at hand. Though important topics are noted in the preceding section, the importance of a managed service to allow flexibility and a way to integrate third-party-specific capabilities is also important when choosing a managed service.

As noted earlier, if an organization chooses to manage their own build solution, they may be responsible for the availability of the platform. In the case of a managed service, the provider is responsible for the availability and may establish an SLA with the customer using its services. The customer would have to make the determination of whether the communicated SLA is acceptable to the business.

Managed services offered by providers reduce the amount of toil to keep the build platform up and running. They allow teams at an organization to focus on critical business functions or revenue-generating activities. In the case of on-premises, not having to wait for hardware procurement or setup allows for maximum business flexibility. The provider would be responsible for making sure the platform is up to date and allowing for fast-paced groups within the organization to experiment with newer capabilities.

Cloud-native automation with Google Cloud Build

This brings us to Cloud Build, a Google Cloud offering that is a serverless platform. This means that teams are not responsible for maintaining, scaling workers, or assuring the availability of the platform. Cloud Build is a managed service as well because Google Cloud is responsible for ensuring that the service is available to the customer. This allows an organization to focus purely on their application code and build pipeline configuration. Google Cloud is responsible for scaling the platform as needed and the billing is priced (`https://cloud.google.com/build/pricing`) per build minute.

Cloud Build takes advantage of the inherent security of Google Cloud, where the API is only accessible by individuals and automation systems that are granted access. This will be a common theme throughout the book because Cloud Build is a serverless platform. We are able to take advantage of what Google Cloud has to offer in areas such as security, compute resources, integrations, and API- and language-specific client SDKs to access the platform.

As noted in previous sections, workers handle the steps (or jobs) in a build pipeline and Cloud Build is responsible for making sure there are workers available to complete the pipeline from end to end. Cloud Build offers workers in different flavors, noted as a default pool and a private pool. In a default pool, the workers are hosted by Google and not visible to the end user. While the build is running, it would have access to the public internet, but not privately hosted resources. Depending on a build that you need, this may be all that is necessary. However, if during a build you need access to resources on a private network, such as dependencies, localized datasets, higher concurrency limits, more machine type configurations, or integration testing with a private environment, the option would be a private pool. Private pools allow for connectivity to private networks using **Google Cloud Platform** (**GCP**) **Virtual Private Cloud** (**VPC**) peering capabilities. The following is a table comparing capabilities of default and private pools:

Feature	Default pool	Private pool
Fully managed	✓	✓
Pay per build minute	✓	✓
Auto scaling, scales to 0	✓	✓
Access to public internet	✓	Configurable
Peer into a VPC or shared VPC to access private resources	⊘	✓
VPC Service Controls support	⊘	✓
Static IP ranges	⊘	✓
Ability to disable public IPs	⊘	✓
Maximum concurrent builds	30	100+
Machine types	5	15
Regions	global	25

Figure 1.4 – Feature comparison between default and private pools

The data in the figure is up to date as of 2022-06-16 (`https://cloud.google.com/build/docs/private-pools/private-pools-overview`).

The steps in a Cloud Build configuration typically remain the same regardless of the type of pool selected: default or private.

In Cloud Build, a single worker is responsible for handling all the steps in a pipeline. The worker and machine type (compute, memory, and storage) can be configured, providing flexibility for each build process, depending on the criticality. Exposing the build process to a machine with more resources may allow for faster builds, where machine resources have been identified as a bottleneck for slower builds.

This is a significant advantage of a serverless platform in Cloud Build, where the options are defined as configuration options. It is the responsibility of Google Cloud to provide the resources based on what is specified in the valid configuration. The flexibility of the cloud is available to the team, by specifying different machine types for the workers based on the pipeline requirements.

If you recall the example of Jenkins (in the *Who needs to manage all of this?* section), it provides plugins to add integrations or functionality to complete a pipeline. In the case of Cloud Build, each individual step consists of a container image known as a **cloud builder**. This provides a significant amount of flexibility as it allows for using Google Cloud-provided builders, community-developed builders, or custom builders, which can be container images created in-house. These custom builders can be any container image that executes a script or binary, which can then be incorporated as a build step.

Organizations may place different restrictions on the residency of code; for example, it may not reside on a developer's workstation, it may only reside on organizationally maintained SCM systems. Cloud Build can integrate with privately hosted/non-GCP solutions, such as GitHub Enterprise. By using private pools, the worker of the build can interact with resources within a private network. This is very important as it allows Cloud Build to support many different scenarios that an organization may have.

Earlier in this chapter, we discussed the value of automation, which includes the automation of build steps, but also how a build itself can be triggered. Triggering a build can be based on a cron schedule using Cloud Scheduler (`https://cloud.google.com/build/docs/automating-builds/create-scheduled-triggers`), async messages through Cloud Pub/Sub (`https://cloud.google.com/build/docs/automating-builds/create-pubsub-triggers`), and webhooks, but is usually driven by a commit of code to an SCM system. Webhooks open the door to many other options; for instance, a build can also be invoked with an API or CLI call through an existing tool. This makes the integration and automation with existing components possible.

GCP service integrations

Integration with existing components is critical, but another advantage of Cloud Build is its native integration with Google Cloud services:

- **Artifact Registry** – store built artifacts (that is, language-specific, container images): `https://cloud.google.com/artifact-registry/docs/configure-cloud-build`.

- **Secret Manager** – natively manage sensitive information: `https://cloud.google.com/build/docs/securing-builds/use-secrets`.

- **VPC Service Controls** – mechanism to protect against data exfiltration, for instance, where a build can be triggered from or policies that require only private pools be used: `https://cloud.google.com/build/docs/private-pools/using-vpc-service-controls`.

- **Binary Authorization** (preview) – sign image attestations at build time with a provided attestor: `https://cloud.google.com/build/docs/securing-builds/use-provenance-and-binary-authorization`.

If you recall earlier, it was noted that a Cloud Build worker is responsible for executing the steps designated in the configured pipeline. The identity of this worker is established by a GCP resource known as a **service account**. Cloud Build has a default service account that it uses, or a user-specified service account may also be specified.

This book focuses on the build aspect of Cloud Build and a built resource does not always need to be deployed when using Cloud Build. However, Cloud Build also provides integrations with instructions and images to deploy built container images to a GCP runtime platform, such as the following:

- Google Kubernetes Engine
- Anthos-enabled clusters
- Cloud Run
- App Engine
- Cloud Functions
- Firebase

Organizations are able to use Cloud Build for both the code build and IaC automation capabilities to provide more autonomy to their teams. One example is noted here: `https://cloud.google.com/customers/loveholidays`.

As a serverless platform, visibility into what is happening with the build is important. Audit logs and build logs are stored by default in GCP's Cloud Logging service. Audit logs cover API calls to the Cloud Build service, such as creation and cancelation. The actual build logs are stored by default in both Cloud Logging and a Google-created Cloud Storage bucket. The logs from Cloud Build are configurable to destinations of your choice within GCP (that is, only Google Cloud Storage, only Cloud Logging, or log buckets in other projects).

Cloud Build, being a serverless platform, also offers an SLA. As with GCP services, Cloud Build has a monthly uptime percentage of 99.95% (at the time of this writing this book – `https://cloud.google.com/build/sla`). This is noted as the **service-level objective (SLO)**. Please refer to the previous link for additional details related to the SLA. For services that are provided by a cloud provider, this is an important measurement factoring in the reliability of the service. As noted in an earlier section, if a team were responsible for managing build infrastructure, they would also need to offer a reliability metric. The built solution would typically have to be designed, implemented, and tested to meet the offered metric. In GCP and its applicable services where SLAs are offered, GCP has this responsibility as a service provider.

Organizations have come a long way from having to manage their own infrastructure and their own pipeline software. Service providers provide a way for the infrastructure to be managed in their data center to offload the hardware, the OS, and sometimes even the management of the build pipeline platform itself. The next iteration is serverless, which is what we've been talking about, where it's just an API that is exposed for clients to use. Define your desired configuration and execute the API in your desired mechanism and language.

Cloud Build is a serverless build pipeline platform that allows organizations to focus on the code at hand. Google Cloud takes care of ensuring the platform is available when needed and that there is sufficient compute for the workers. Each step, represented as a container image, provides the utmost flexibility in handling pipeline needs.

Summary

Thus far, we've covered how build pipeline automation can make our lives easier while also potentially preventing error-prone mistakes. Metadata and metrics can help us make data-driven decisions on making our pipelines more efficient. The build platforms need to be managed by someone, whether it's an IT team, an outsourced team, or a service managed by a cloud provider. If someone else manages it, it may allow us to focus more on our business needs rather than pipeline platform needs. Cloud Build is a serverless platform providing us with worker pool elasticity and flexibility using container images as build steps.

Stay tuned for further chapters as we dig into how to get started and deeper details on the features of Cloud Build and run through examples of a few scenarios.

In the next chapter, we will focus on getting our Cloud Build environment set up.

2
Configuring Cloud Build Workers

Having discussed a little about the context of automation in the industry and where **Cloud Build** fits into this, let's dive into the implementation details of Cloud Build. We will start with Cloud Build workers—the machines that execute your builds.

The possibilities for what builds can automate and accomplish are numerous. They can define a series of steps that test source code and translate it into an executable binary. They can define a number of cloud infrastructure resources to be created, updated, or torn down. They can read from and write to various external systems, such as Git-based repositories or container registries. They can define workflows for data processing or **machine learning** (**ML**) pipelines.

When writing builds for Cloud Build, each step runs in a container, and that container image can be built and provided by you. As we discussed in the previous chapter, these container images that run a build step are called **builders**.

Given that the use cases for builds are many, it is important to understand how to configure the environment in which those builds run. These builds may have certain requirements for their execution environment, including the following:

- Specific amounts of **central processing units** (**CPUs**) and memory resources
- Network connectivity to resources running on a private network
- Locality to a specific region due to compliance or latency necessities

This execution environment runs on **virtual machines** (**VMs**) that are fully managed by Google, otherwise known as **worker pools**.

In this chapter, we will review the architecture of Cloud Build and the various ways you can configure workers to execute your builds, walking away with knowledge of the following topics:

- How worker pools can be configured in Cloud Build
- Prerequisites for running builds on worker pools
- Using the default pool
- Using private pools

Technical requirements

- Data center and infrastructure concepts
- Public cloud concepts
- Software build concepts
- Networking concepts

How worker pools can be configured in Cloud Build

The term *managed service* has done some heavy lifting in the software industry, with the term yielding varying definitions of what *managed* means, depending on the service. Thus, it is important when using any managed service to review the architecture where the boundaries lie between the provider and you, the user. Cloud Build is no different!

In Cloud Build, the architecture of where workers run depends on which type of worker pool your build executes in. There are two types of worker pools in Cloud Build, as outlined here:

- The **default pool**, which is the simplest way to run builds on fully managed workers
- **Private pools,** which are fully managed workers configured to a higher level, including connectivity to private **Virtual Private Cloud** (**VPC**) networks

Despite having various types of worker pools supported by Cloud Build, it is key to call out that the core user journey of having builds run in Cloud Build aims to be consistent regardless of which type of worker pool you utilize, for both authoring build definition and submitting build definitions to be executed. In the following diagram, we can see that the Cloud Build **application programming interface** (**API**) should be consistent across all worker pools:

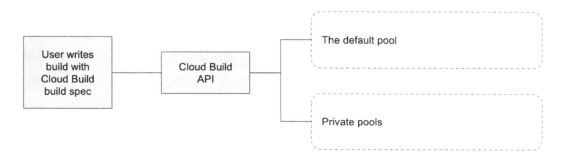

Figure 2.1 – Builds in Cloud Build can run on various worker pool types with minimal-to-no changes

This means that regardless of which type of worker pool executes your build, the method in which you specify what a build looks like remains the same. We will review in depth Cloud Build's build specification in later chapters, but let's first discuss the minimum requirements in a **Google Cloud Platform (GCP)** environment to get started with Cloud Build.

Prerequisites for running builds on worker pools

When beginning to work with Cloud Build, it's important to understand how its resource hierarchy works in your Google Cloud environment. Cloud Build and all the various worker pools we will discuss are scoped to a **Project**.

Your Google Cloud project will be the top-level resource that contains resources that you will utilize when using Cloud Build. Within a Project, you can utilize **Identity and Access Management (IAM)** to create user accounts and service accounts that will grant you and others access to various resources within the Project. To ensure that you can proceed with Cloud Build and the following examples, ensure that you have the following roles associated with your user in IAM:

- **Cloud Build Editor**
- **Network Admin**
- **Logging Admin**

These roles will enable you to utilize Cloud Build and the services it integrates with in Google Cloud, as follows:

- Cloud Build, for creating builds
- VPC, for network connectivity for certain worker pools
- Cloud Logging, for accessing build logs

This is recommended for a sandbox environment only. In later chapters, we will review how to utilize IAM for Cloud Build while applying the **principle of least privilege**.

The resources used within a Project with Cloud Build are summarized in the following diagram:

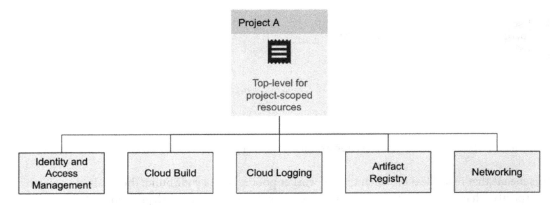

Figure 2.2 – Project-scoped services typically utilized when using Cloud Build

In later chapters, we will discuss how to apply the principle of least privilege to Cloud Build and its associated service, but for now, let's continue forward.

In order to be able to spin up resources beyond Google Cloud's free tier, you will also need a **billing account** associated with your project. This account can be associated with more than a single project. Costs for Cloud Build can be found at `https://cloud.google.com/build/pricing`. The pricing will vary depending on the machine types and worker pool being utilized; however, this generally follows the model of **United States dollars** (**USD**) $/build minute, in line with the expectation that these resources spin up and down dynamically based on the volume of builds running. The following diagram shows the relationship between projects and billing accounts:

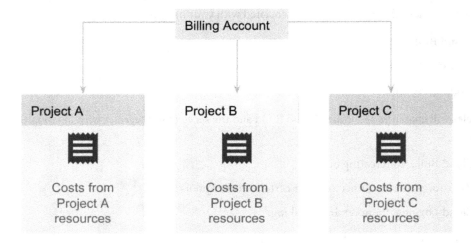

Figure 2.3 – Relationship between projects and billing accounts

Once you have a project, IAM, and a billing account sorted out, you will then need to enable the Cloud Build and Service Networking APIs in your Project.

For this and all following examples within this book, we will be utilizing Cloud Shell, a free and ephemeral developer workstation provided by Google Cloud. It runs a Debian machine with multiple tools already installed, including gcloud, which is the **command-line interface (CLI)** for utilizing Google Cloud services.

Open Cloud Shell (https://shell.cloud.google.com/) and authenticate with gcloud by running the following command and following the instructions in the response:

```
$ gcloud auth login
```

Once authenticated, run the following command to enable the required APIs to use Cloud Build and configure networking for private pool workers:

```
$ gcloud services enable cloudbuild.googleapis.com
servicenetworking.googleapis.com
```

With your Google Cloud project set up, let's now review the various ways worker pools can be configured!

Using the default pool

At its initial **General Availability (GA)** launch in 2017, Cloud Build was designed to provide users with a simple interface and fully managed experience in Google Cloud when building container images and other artifacts such as Go executables or Java archives.

> Note
> Cloud Build was initially launched under the name **Container Builder** in 2017, underscoring its simple and focused purpose to provide end users with a way to automate builds.

From its inception, Cloud Build has aimed to ensure the following:

- Users could create builds and triggers for builds by interacting with Google's API or **user interface (UI)**.
- Users would not need to create, patch, scale, or manage the worker machines that run builds.
- Users would have pre-built integration with logging and monitoring to introspect builds.

Cloud Build first accomplished this experience by providing users with the first mode for workers in Cloud Build: **the default pool**.

The default pool is naturally named as it is the default method for Cloud Build to provision workers to execute your builds. When submitting a build to Cloud Build, if you do not specify which type of worker pool should execute your build, it will be run in the default pool. This pool is available for you to use as soon as you enable the Cloud Build API and have permissions to execute builds.

This mode provides the simplest **user experience (UX)** among the various ways you can utilize Cloud Build. Consider the following flow diagram for when a user creates a build to run on a worker in the default pool:

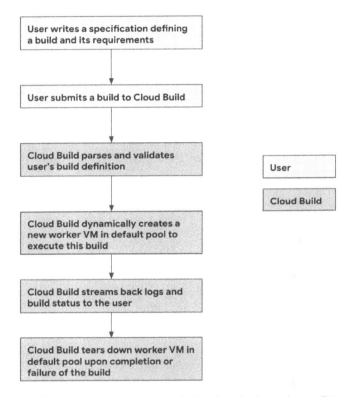

Figure 2.4 – Delineation between what users do and what Google does when utilizing the default pool

The user is primarily focused on their specific build steps and requirements, which ensures their time is spent only on the automation they seek to achieve and not on managing any of the infrastructure that will run their build.

Cloud Build enables this experience by ensuring that all of the infrastructure provisioned is created in a Google-owned environment where Google's engineers can own the provisioning, securing, and management of the underlying build infrastructure.

Cloud Build provisions one dedicated worker for each build, created with a machine type according to that build's specification for resource requirements. If no machine type is specified, the default machine type will be used, which utilizes one **virtual CPU (vCPU)** and 4 **gigabytes (GB)** of memory.

Note

Cloud Build has quotas and limits set at the project level around numerous dimensions of a build.

One of the most important limits is around concurrency, which is the number of builds that can be running at the same time. For the default pool, concurrency is set to 30 concurrent builds in a given project. For private pools, you can have up to 100 concurrent builds in a given private pool, and up to 10 private pools in a given project. This means that private pools have a much higher concurrency limit for running numerous builds simultaneously.

This is based on the default limitations set in a project; you can find a full list of quotas, limits, and possible exceptions here: `https://cloud.google.com/build/quotas`.

The default pool will have as many workers as there are builds running, ensuring each build has an isolated environment in which it can execute. If many builds are running at the same time with resource requirements, the single default pool may be running workers with multiple machine types. The following diagram shows the default pool architecture:

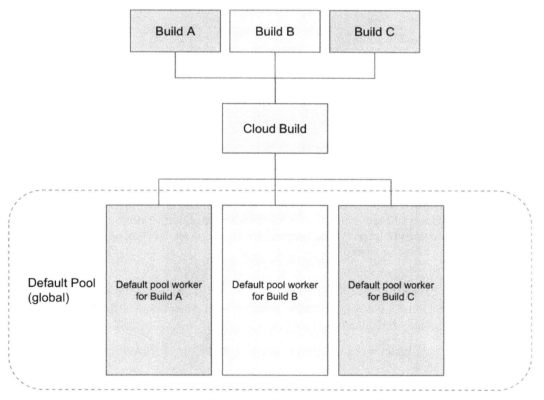

Figure 2.5 – Default pool architecture

Each worker is also provisioned with outbound access to the public internet. When you submit builds to execute in the default pool, you can configure the following:

- CPU and memory to be utilized by the build
- Storage to be available to all of the build steps in the build

You assign CPU and memory resources by specifying a machine type for the worker. Each machine type has a predetermined amount of CPU and memory; you can choose between these machine types based on the computing requirements of your build. The following machine types are available in the default pool, at publication time:

Machine type	vCPUs	Memory (GB)	Warm start	Default
E2-medium	1	4	Yes	Yes
E2-highcpu-8	8	8	No	No
E2-highcpu-32	32	32	No	No
N1-highcpu-8	8	8	No	No
N1-highcpu-32	32	32	No	No

Table 2.1 – Default pool machine types and their respective resources

As for storage, each worker by default has 100 GB in network-attached **solid-state drive (SSD)** storage. Users can specify up to an additional 1,000 GB of network-attached SSD storage in their build configuration, which we will review in subsequent chapters.

> **Note**
> These workers do not have any access to your resources on machines within your private VPC network. With that said, Cloud Build workers do run with a specified service account that can be granted permissions to interact with resources in your project via Google APIs, such as managed artifact stores (for example, Container Registry and Artifact Registry).

To begin running builds on the Cloud Build default pool, there is no preparation required for users to create any resources. Rather, users merely need to submit the build definition to Cloud Build, which we will review in subsequent chapters.

Now, let's move on to the concept of private pools in Cloud Build.

Using private pools

Private pools are similar to the default pool in that they are fully managed workers that live in a Google-managed environment. Designed for more complex sets of requirements from those looking to run fully managed builds, private pools help users who require the following:

- To run resources on a private VPC network that must be accessed from a build
- More optionality around machine types, such as high-memory machines
- To further secure the network perimeter of their build environment

Unlike the default pool, you can have multiple private pools in a single Google Cloud project. You create empty private pool resources that contain a definition of what a private pool worker should look like, with details such as their machine type and the network they should connect to. Google will then scale up these private pools from zero workers when builds are scheduled to run on them.

With that said, the workflow for end users submitting builds remains consistent, as we can see here:

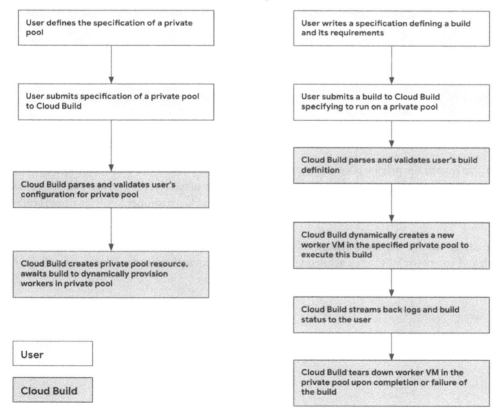

Figure 2.6 – Delineation between what users do and what Cloud Build does when using private pools

The architecture itself is slightly different from the default pool. As previously discussed, you can have multiple private pools in a single project, each with various configurations. What does remain the same is that a build is isolated to a single worker in the private pool it runs in. The following diagram shows the private pool architecture:

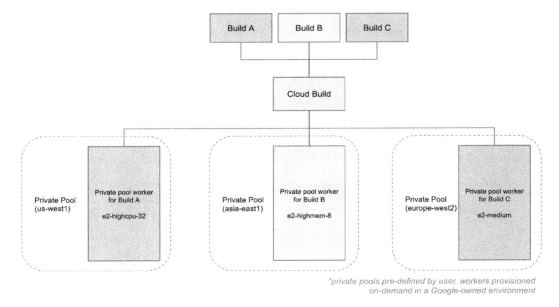

Figure 2.7 – Private pool architecture

One unique trait of the private pool architecture is that private pool workers can have connectivity to your own VPC networks in your Google Cloud Project. This network connectivity is made possible by the architecture of private pools, in which the Google-owned environment utilizes what is called a **Service Producer network**, where these workers are assigned internal **Internet Protocol** (**IP**) addresses.

You can utilize the **Service Networking API** to then create a peering between that Service Producer network and your own VPC, thus achieving network connectivity between workers and your private VPC-scoped resources, as illustrated in the following diagram:

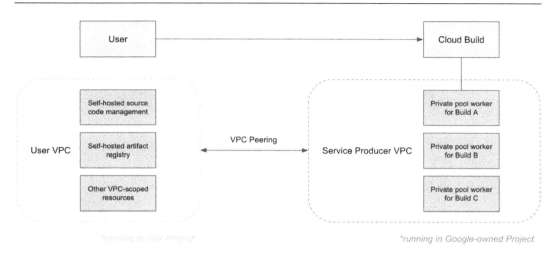

Figure 2.8 – Network architecture for private pools

With this networking functionality, not only can you access resources within your VPC, but also source code management systems and artifact registries that may reside on-premises connected to your VPC by a private connection such as a **Virtual Private Network** (**VPN**) tunnel or a **Direct Interconnect**.

In addition to configurable networking, private pools allow users to specify a specific Google Cloud region in which their private pool and workers will reside. This is in contrast to the default pool, which executes on a worker that can run in multiple regions around the globe, depending on resource availability.

Private pools can be created in the following regions, at the time of publication:

- Asia:
  - asia-east1 (Taiwan)
  - asia-east2 (Hong Kong)
  - asia-northeast1 (Tokyo)
  - asia-northeast2 (Osaka)
  - asia-northeast3 (Seoul)
  - asia-south1 (Mumbai)
  - asia-south2 (Delhi)
  - asia-southeast1 (Singapore)
  - asia-southeast2 (Jakarta)

- Australia:

  - australia-southeast1 (Sydney)

  - australia-southeast2 (Melbourne)

- Europe:

  - europe-central2 (Warsaw)

  - europe-north1 (Finland)

  - europe-west1 (Belgium)

  - europe-west2 (London)

  - europe-west3 (Frankfurt)

  - europe-west4 (Netherlands)

  - europe-west6 (Zurich)

- North America:

  - northamerica-northeast1 (Quebec)

  - northamerica-northeast2 (Toronto)

  - us-central1 (Iowa)

  - us-east1 (South Carolina)

  - us-east4 (Virginia)

  - us-west1 (Oregon)

  - us-west2 (Los Angeles)

  - us-west3 (Salt Lake City)

  - us-west4 (Las Vegas)

- South America:

  - southamerica-east1 (Sao Paolo)

  - southamerica-west1 (Santiago)

Private pools also expand the available sets of machine types that you can specify for workers. The machine types available are as shown here:

Machine type	vCPUs	Memory (GB)	Warm start	Default
E2-medium	1	4	No	Yes
E2-standard-2	2	8	No	No
E2-standard-4	4	16	No	No
E2-standard-8	8	32	No	No
E2-standard-16	16	64	No	No
E2-standard-32	32	128	No	No
E2-highmem-2	2	16	No	No
E2-highmem-4	4	32	No	No
E2-highmem-8	8	64	No	No
E2-highmem-16	16	128	No	No
E2-highmem-32	32	256	No	No
E2-highcpu-2	2	2	No	No
E2-highcpu-4	4	4	No	No
E2-highcpu-8	8	8	No	No
E2-highcpu-16	16	16	No	No
E2-highcpu-32	32	32	No	No

Table 2.2 – Private pool machine types and their respective resources

While private pools maintain the same fully managed experience that the default pool provides, a private pool must be created in advance of actual build execution. Let's review how we might do this by creating a private pool.

First, in order to demonstrate some of the networking capabilities that private pools provide, we should create a sandbox VPC that we can use to peer with the Google-owned environment hosting your private pool. Here's the code we'll need in order to achieve this:

```
$ project_id=$(gcloud config get-value project)

$ vpc_name=packt-cloudbuild-sandbox-vpc

$ vpc_range_name=packt-cloudbuild-sandbox-vpc-peering-range

$ gcloud compute networks create $vpc_name --subnet-mode=custom
```

Once a VPC is created, next, create an IP range that does not overlap with any IP space in your VPC that can be used for the internal IPs of workers in your private pool.

In the following example, you'll be utilizing the 192.168.0.0/16 space, as defined in the addresses and prefix-length arguments. Note that the size of your range will determine how many workers you can have in this private pool:

```
$ gcloud compute addresses create packt-cloudbuild-sandbox-vpc-
peering-range \
        --global \
        --purpose=VPC_PEERING \
        --addresses=192.168.0.0 \
        --prefix-length=16 \
        --description="IP range for private pool workers" \
        --network=$vpc_name
```

Finally, create VPC peering with the following gcloud command; this will reference that you want to peer your VPC with the service producer network VPC in which the private pool workers reside:

```
$ gcloud services vpc-peerings connect \
      --service=servicenetworking.googleapis.com \
      --ranges=$vpc_range_name \
      --network=$vpc_name \
      --project=$project_id
```

Now that we've configured a VPC that our private pool can connect to, let's create a private pool itself. There are a few ways to do this currently, as outlined here:

- Via the Cloud console
- Via passing a **YAML Ain't Markup Language** (**YAML**)-based configuration file to gcloud
- Via passing the configuration directly to gcloud

Some of the configurations available for private pools are similar to what you are able to configure in default pools—namely, the following:

- Machine type for each worker, albeit with more available machine types
- Network-attached SSD for each worker, up to 1,000 GB

However, you'll notice a few new configuration options as well, as outlined here:

- Peered network, to specify which VPC the workers in the private pool will peer with
- Region, as a private pool and the workers in it are scoped to a single region
- Public egress, as you can specify whether or not workers in a private pool have public IP addresses

Let's now demonstrate some of the functionality of private pools by creating a private pool using gcloud while passing the configuration directly. In this case, we will be creating a private pool of workers that have the e2-standard-2 machine type, with 100 GB of network-attached SSD, and located in us-west1. The code is illustrated in the following snippet:

```
$ gcloud builds worker-pools create packt-cloudbuild-
privatepool \
    --project=$project_id \
    --region=us-west1 \
    --peered-network=projects/$project_id/global/networks/$vpc_
name \
    --worker-machine-type=e2-standard-2 \
    --worker-disk-size=100GB \
    --no-public-egress
```

Once we have a private pool actually created, we can view it in the Cloud console, as shown in the following screenshot:

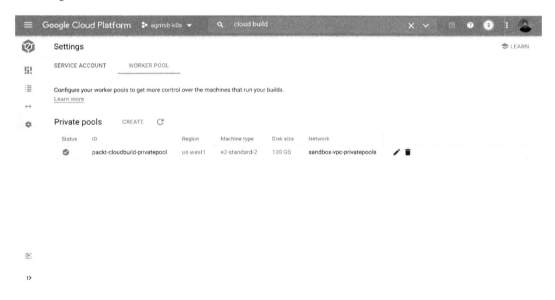

Figure 2.9 – Viewing the previously created private pool in the Cloud console

Once we have this created, we can begin to run builds on workers defined by the private pool.

Summary

In this chapter, we have discussed how to get started with enabling Cloud Build in your Google Cloud project and how to navigate the multiple ways in which you can create workers to execute your builds.

Specifically, we have reviewed the architecture of the default pool and private pools. We have also discussed the user requirements for creating private pools.

In the next chapter, we will proceed with reviewing the specification for builds in Cloud Build and run your first build!

3

Getting Started – Which Build Information Is Available to Me?

As a managed service, Cloud Build allows individuals and organizations to focus on the steps in the build pipeline. Even though it's managed, it's good to understand what happens after a build is started and which resources may be available to track status and view progress via logs through automated means or a **graphical user interface** (GUI) such as the Google Cloud console.

In this chapter, we will cover the following topics:

- How your build resources are accessed
- Build submission and status
- Using the **Google Cloud Platform** (GCP) console
- Build operations

Technical requirements

- Software build concepts and tools
- GCP console familiarity

How your build resources are accessed

In order to create reproducible and consistent builds, the files and resources used during a build should be referenceable. The quickest and easiest way is when a **source control management** (SCM) tool is used, it can be referenced by a `commit` tag that is immutable. This allows us to essentially get a snapshot of the files used during a build so that if it were needed to be run again, we have at least a constant in place.

Cloud Build needs a way to fetch the build files, and this can be from the SCM tool. However, what if the build files are located on the filesystem and the build was initiated by the `gcloud` **command-line interface** (**CLI**) command? In this situation, `gcloud` will package up the files in the specified folder, along with subfolders if present. The files are packaged in a tarball (gzipped archive) and uploaded to **Google Cloud Storage** (**GCS**). Cloud Build will then fetch the packaged files from GCS and begin the build.

The `gcloud` command also gives users the flexibility to determine which files should be included in a build, through a `.gcloudignore` file that uses the same syntax as `gitignore` (`https://git-scm.com/docs/gitignore`). If a `.gcloudignore` file is not created, `gcloud` is able to generate one based on a `.gitignore` file in the top-level folder.

Whether it's through an SCM that can provide a snapshot of files or snapshots packaged by `gcloud` and uploaded to GCS, we will always have a reference point for each build that is run. As noted earlier, this is very important for tracing back what was included in a build for auditing and debugging purposes. The Cloud Build `gcloud` command also allows users to specify a gzipped archive manually if they have some local build automation.

Build submission and status

Cloud Build as a managed service provides multiple methods to initiate a build to enable maximum flexibility. For instance, it can be triggered in the following ways:

- Manually
- Automated through other tools
- Messages published to **Publish/Subscribe** (**Pub/Sub**)
- On a schedule
- Changes pushed to the SCM system

A trigger is a mechanism in Cloud Build to start a build pipeline; for instance, this can be a change to source code in an SCM system or an event from another system. Triggers will be covered in more depth in *Chapter 5, Triggering Builds*.

Note

The commands and output in this chapter are for illustrative purposes to describe capability as well as analysis of the output where applicable. Some of these commands may require prerequisites to be configured in order to execute properly.

First, we start off with a simple `cloudbuild.yaml` file with two steps that build a Docker image and push it to Artifact Registry. The following example is a simplified version of the `cloudbuild.yaml` file. A snippet can be found here: https://github.com/GoogleCloudPlatform/cloud-build-samples/blob/main/basic-config/cloudbuild.yaml. Please note the contents in GitHub may change over time:

```
# Docker Build
- name: 'gcr.io/cloud-builders/docker'
  args: ['build', '-t',
         'us-central1-docker.pkg.dev/${PROJECT_ID}/image-
repo/myimage',
         '.']

# Docker Push
- name: 'gcr.io/cloud-builders/docker'
  args: ['push',
         'us-central1-docker.pkg.dev/${PROJECT_ID}/image-
repo/myimage']
```

In this chapter, we will be focusing on *manual* methods, as other automated mechanisms will be covered in later chapters. To get us started, the quickest way to initiate a build is through the `gcloud` CLI tool. Cloud Build, as a managed service providing an automation platform, is able to abstract out the need for managing resources and infrastructure for the build to execute. As noted in *Chapter 2, Configuring Cloud Build Workers*, private pools would require some additional configuration, but the steps in the configuration largely remain the same. The following command will submit a build and include the contents of the current path and subpaths:

```
$ gcloud builds submit
```

Here is a sample output of the preceding command:

```
[1] Creating temporary tarball archive of 2 file(s) totalling
952 bytes before compression.
Uploading tarball of [.] to [gs://##project-id_redacted##_
cloudbuild/source/1641339906.759161-f251af6f4c9745438a4730c3c6
f94cd0.tgz]
[2] Created [https://cloudbuild.googleapis.com/v1/
projects/##project-id_redacted##/locations/global/builds/
ca6fd20f-7da3-447e-a213-7b542f9edb5c].
[3] Logs are available at [https://console.cloud.
google.com/cloud-build/builds/ca6fd20f-7da3-447e-a213-
```

```
7b542f9edb5c?project=##project-number_redacted##].
-------------------------------------------- REMOTE BUILD
OUTPUT ----------------------------------------------------
starting build "ca6fd20f-7da3-447e-a213-7b542f9edb5c"

FETCHSOURCE
[4] Fetching storage object: gs://##project-id_redacted##_
cloudbuild/source/1641339906.759161-f251af6f4c9745438a4730c3c6f
94cd0.tgz#1641339907144359
Copying gs://##project-id_redacted##_cloudbuild/
source/1641339906.759161-f251af6f4c9745438a4730c3c6f94cd0.
tgz#1641339907144359...
/ [1 files][  695.0 B/  695.0 B]
Operation completed over 1 objects/695.0 B.
BUILD
[5] Starting Step #0
...
Step #0: Successfully tagged us-central1-docker.pkg.
dev/##project-id_redacted##/image-repo/myimage:latest
Finished Step #0
Starting Step #1
Step #1: Already have image (with digest): gcr.io/cloud-
builders/docker
...
Finished Step #1
PUSH
DONE
-----------------------------------------------------------
------------------------------------------------------
ID: ca6fd20f-7da3-447e-a213-7b542f9edb5c
CREATE_TIME: 2022-01-04T23:45:07+00:00
DURATION: 9S
SOURCE: gs://##project-id_redacted##_cloudbuild/
source/1641339906.759161-f251af6f4c9745438a4730c3c6f94cd0.tgz
IMAGES: -
[6] STATUS: SUCCESS
```

The preceding output is now outlined in further detail (note that the preceding numbers are added for illustration purposes; they are not in the command output):

[1] The output where a temporary tarball archive was created and then uploaded to GCS. This serves two purposes, as described here:

- Allows for the build files to be fetched by the Cloud Build service
- Provides a snapshot of the build files for auditing and debugging

[2] A Cloud Build **globally unique identifier** (**GUID**) uniquely identifying the build is also generated in link form to allow for access to the build from the Google Cloud console UI.

[3] A link to where the build logs can be accessed from the Google Cloud console UI.

[4] Cloud Build fetches the archive uploaded to GCS by gcloud.

[5] The build begins with the first step specified in the cloudbuild.yaml file.

[6] The final status of the build.

When invoking a build manually with the CLI, by default it is performed synchronously, which means that the output of the build is locking on the terminal. The terminal prompt will not be returned until the build finishes successfully or with a failure. If the user chooses to break out of the build (by pressing *Ctrl + C*), this will also cancel the build. Alternatively, you can pass in the --async option in order to queue the build and return the user to the prompt. The following command will list builds along with their status:

```
$ gcloud builds list
```

Here is a sample output of the preceding command:

```
ID: ca6fd20f-7da3-447e-a213-7b542f9edb5c
CREATE_TIME: 2022-01-04T23:45:07+00:00
DURATION: 9S
SOURCE: gs://##project-id_redacted##_cloudbuild/
source/1641339906.759161-f251af6f4c9745438a4730c3c6f94cd0.tgz
IMAGES: -
STATUS: SUCCESS

ID: 2522e42e-c31f-4b7a-98bc-404b441a5f2c
CREATE_TIME: 2022-01-04T23:38:44+00:00
DURATION: 1M50S
SOURCE: gs://##project-id_redacted##_cloudbuild/
source/1641339523.810603-97be6bc031b44b83bded5a6b0133e093.tgz
```

```
IMAGES: -
STATUS: FAILURE
```

The preceding sample output provides a brief summary of each build within the project. Using the build ID from either the build submission or the list, we are able to describe the build in detail, as illustrated here:

```
$ gcloud builds describe ca6fd20f-7da3-447e-a213-7b542f9edb5c
```

Here is a sample output of the preceding command:

```
createTime: '2022-01-04T23:45:07.383980674Z'
finishTime: '2022-01-04T23:45:17.496050Z'
id: ca6fd20f-7da3-447e-a213-7b542f9edb5c
logUrl: https://console.cloud.google.com/cloud-build/builds/
ca6fd20f-7da3-447e-a213-7b542f9edb5c?project=##project-number_
redacted##
logsBucket: gs://##project-number_redacted##.cloudbuild-logs.
googleusercontent.com
name: projects/##project-number_redacted##/locations/global/
builds/ca6fd20f-7da3-447e-a213-7b542f9edb5c
...
startTime: '2022-01-04T23:45:08.020795233Z'
status: SUCCESS
steps:
- args:
  - build
  - -t
  - us-central1-docker.pkg.dev/##project-id_redacted##/image-
repo/myimage
  - .
  name: gcr.io/cloud-builders/docker
  pullTiming:
    endTime: '2022-01-04T23:45:13.814938454Z'
    startTime: '2022-01-04T23:45:13.811085177Z'
  status: SUCCESS
  timing:
    endTime: '2022-01-04T23:45:15.114544455Z'
    startTime: '2022-01-04T23:45:13.811085177Z'
```

```
...
timeout: 600s
timing:
  BUILD:
    endTime: '2022-01-04T23:45:17.020766002Z'
    startTime: '2022-01-04T23:45:13.227178424Z'
  FETCHSOURCE:
    endTime: '2022-01-04T23:45:13.227099439Z'
    startTime: '2022-01-04T23:45:08.417349649Z'
```

The output from the build description provides detail on the build as well as details of each step, such as start and stop time. As part of gcloud, the output can be output in other formats such as **JavaScript Object Notation (JSON)** for querying and aggregating the output for multiple builds.

There are many ways to interact with Cloud Build. The scenario using gcloud was used in the preceding example to interact with Cloud Build, but there are additional methods as well, as outlined here:

- **REpresentational State Transfer (REST) application programming interfaces (API)** (https://cloud.google.com/build/docs/api/reference/rest)

- Client libraries (https://cloud.google.com/build/docs/api/reference/libraries)

This will help provide flexibility when trying to integrate with existing tooling, infrastructure, and developer/operator preferences. Cloud Build can be the **end-to-end (E2E)** pipeline or just be involved in specific steps where desired.

The simplicity of a standard schema for defining pipeline steps across different worker platforms is key to allowing organizations to focus more on their business applications and services, rather than the automation platform itself. Once a build is submitted, it's also very important to quickly get the status of a build and audit and troubleshoot if necessary. Additional auditing metadata is also available for builds that push Docker images to Artifact Registry. The metadata is known as build provenance, which can contain the source code repository, build steps, and the image path. While it is metadata, it also contains signature data that can be verified against the Cloud Build public key for integrity, and this helps ensure the metadata has not been manipulated. This content will be covered in a later chapter. Just because a service is managed, this doesn't mean it should be a black box when needing to integrate or solve issues.

Using the GCP console

Interacting with the CLI or APIs is very important for automation, but for certain scenarios, it's just as important to be able to quickly get a high-level summary in an organized way on a UI. Cloud Build also has UI capabilities to perform some operations instead of interacting with the API.

The Cloud Build summary **Dashboard** section provides the following information:

- A summary of triggers
- The latest build of the respective trigger
- Build history filtered by the trigger
- The average duration of builds
- The percentage of builds passing and failing, based on the last 20 builds

If a project has a lot of configured triggers or there is a desire to show certain items, the listing of triggers can also be filtered by the following:

- Trigger source
- Trigger name
- Trigger description

A pin may also be placed on triggers that should be shown toward the top of the list. For each trigger, if the build configuration contains a step for approval or rejection, this can also be performed on the **Dashboard** view on the latest build of a particular trigger, as illustrated in the following screenshot:

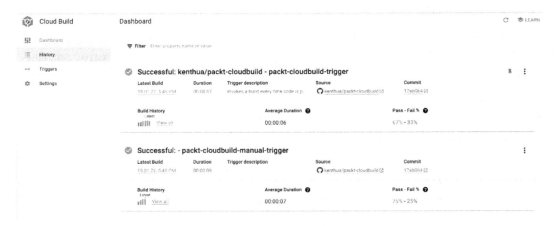

Figure 3.1 – Sample dashboard

A listing of each build may be accessed through the **History** section, where columns can be customized to show the desired information. Filters can also be used to show only specific builds with specific metadata, as illustrated in the following screenshot:

Figure 3.2 – Sample history

Triggers are a critical component for automating builds in Cloud Build. This section of the UI allows for the creation of triggers and setting up SCM repositories that will be utilized by the trigger, as illustrated in the following screenshot:

Figure 3.3 – Sample triggers

Cloud Build supports many SCM offerings, from hosted public repositories to private repositories such as GitHub Enterprise. You can see an example of this in the following screenshot:

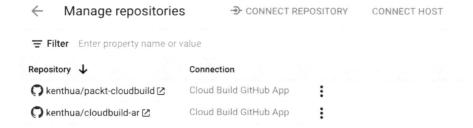

Figure 3.4 – Sample triggers: managing repositories

Additional settings for the Cloud Build managed service allow for permission settings and the management of private pools.

Specific to permissions, Cloud Build utilizes a default service account, and within this view, a user is able to set permissions for specific cloud services, ranging from compute runtimes to secret stores. You can see an example of this in the following screenshot:

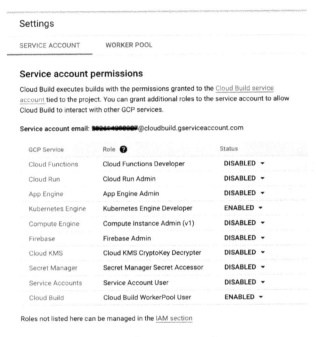

Figure 3.5 – Sample settings: service account

The creation and details of private pools are made available over the UI as well, as illustrated in the following screenshot:

Figure 3.6 – Sample settings: worker pool

The GCP console UI provides another way for users to interact with the Cloud Build managed service. For users just getting started or those that only have a few builds, it may make sense to only integrate with the UI. However, human error and repetition can be reduced by utilizing the CLI and/or API integrations.

Build operations

Congratulations—we've submitted a build via the CLI and navigated across the UI for some additional details! Operationally, there are quite a few things we may want to do after a build has been kicked off. A few thoughts come to mind, and these are set out here:

- View the logs
- View the build history
- Rebuild
- View the audit logs
- Build provenance

Let's get started with how to access the logs of a build that has already been kicked off. We'll continue the example using `gcloud`. Leveraging the ID of the build, we are also able to retrieve the logs for the build, like so:

```
$ gcloud builds log ca6fd20f-7da3-447e-a213-7b542f9edb5c
```

Here is a sample output of the preceding command:

```
----------------------------------------------- REMOTE BUILD
OUTPUT ------------------------------------------------
starting build "ca6fd20f-7da3-447e-a213-7b542f9edb5c"

FETCHSOURCE
Fetching storage object: gs://##project-id_redacted##_
cloudbuild/source/1641339906.759161-f251af6f4c9745438a4730c3c6f
94cd0.tgz#1641339907144359
Copying gs://##project-id_redacted##_cloudbuild/
source/1641339906.759161-f251af6f4c9745438a4730c3c6f94cd0.
tgz#1641339907144359...
/ [1 files][  695.0 B/  695.0 B]
Operation completed over 1 objects/695.0 B.
BUILD
Starting Step #0
```

```
Step #0: Already have image (with digest): gcr.io/cloud-
builders/docker
Step #0: Sending build context to Docker daemon  3.584kB
Step #0: Step 1/2 : FROM alpine
Step #0: latest: Pulling from library/alpine
Step #0: Digest: sha256:21a3deaa0d32a8057914f36584b5
288d2e5ecc984380bc0118285c70fa8c9300
Step #0: Status: Downloaded newer image for alpine:latest
Step #0:  c059bfaa849c
Step #0: Step 2/2 : CMD ["echo", "Hello World!"]
Step #0:  Running in 35337ee2efec
Step #0: Removing intermediate container 35337ee2efec
Step #0:  1071be434322
Step #0: Successfully built 1071be434322
Step #0: Successfully tagged us-central1-docker.pkg.
dev/##project-id_redacted##/image-repo/myimage:latest
Finished Step #0
Starting Step #1
. . .
Finished Step #1
PUSH
DONE
```

The log output of each step in the build is helpful for understanding what took place during a build but also for debugging if necessary. Understanding what takes place, in conjunction with details around how long a step takes, can also help drive the optimization of pipeline steps. If a build was initiated asynchronously with the --async option, we can use the --stream option for the log command to be able to view log output as it happens.

Logs can also be viewed using the GCP console UI for Cloud Build, from the build history. For each build, we can see logs for the complete build, as illustrated in the following screenshot:

Figure 3.7 – Sample overall build log output via the GCP console UI: Cloud Build view

Logs for a specific build step within the build can be accessed by selecting an individual build step, as illustrated in the following screenshot:

Figure 3.8 – Sample single build step log output via the GCP console UI: Cloud Build view

Raw logs can also be viewed through the GCP console UI Cloud Logging service. Notice in the following screenshot that each log entry consists of a `textPayload` of log data but also metadata associating the text with the build (such as `build_id`):

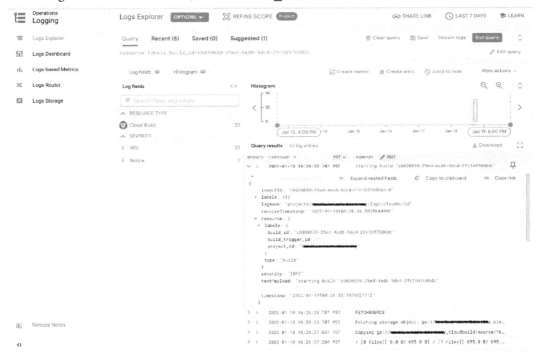

Figure 3.9 – Sample overall build log output via the GCP console UI: Cloud Logging view

Rebuilding is also available through the GCP console UI; it is based on a specific commit or source artifact. Initiating a rebuild will create a new build with a new build ID, using the original source artifacts. If a source artifact causes a build to fail, it's likely rebuilding will also continue to fail because it's using the same resources. However, if a build is dependent on an external system during integration testing or separate source files as part of a build, this may cause a build to result in different behavior.

Audit logs are also critical in understanding who performed certain actions on builds, such as creation, approvals, updates, and who accessed build metadata. Audit logs are available through the Cloud Logging service. The Cloud Logging service is available on the GCP console as well as through APIs (for example, REST API; **Software Development Kit** (**SDK**); the CLI). Here is an example of using the `gcloud` tool via the CLI to view audit logs:

```
$ gcloud logging read "logName:cloudaudit.googleapis.
com%2Factivity AND \
    resource.type=build"
```

Here is a sample output of the preceding command:

```
insertId: 166bo1hd1dvc
logName: projects/##project-id_redacted##/logs/cloudaudit.
googleapis.com%2Factivity
operation:
  id: operations/build/##project-id_redacted##/
YzYwMzA2NTktMjVlZC00ZWQ2LTlkYzQtMmZjMWI5NzY4NmRj
  last: true
  producer: cloudbuild.googleapis.com
protoPayload:
  '@type': type.googleapis.com/google.cloud.audit.AuditLog
  authenticationInfo:
    principalEmail: ##email_redacted##
  authorizationInfo:
  - granted: true
    permission: cloudbuild.builds.create
    resource: projects/00000041cd1ce997
    resourceAttributes: {}
  methodName: google.devtools.cloudbuild.v1.CloudBuild.
CreateBuild
  requestMetadata:
    destinationAttributes: {}
    requestAttributes: {}
  resourceName: projects/##project-id_redacted##/builds
  serviceName: cloudbuild.googleapis.com
  status: {}
receiveTimestamp: '2022-01-19T00:26:46.102748864Z'
resource:
  labels:
    build_id: c6030659-25ed-4ed6-9dc4-2fc1b97686dc
    build_trigger_id: ''
    project_id: ##project-id_redacted##
  type: build
severity: NOTICE
timestamp: '2022-01-19T00:26:45.655472495Z'
```

While the preceding text is output from the CLI, the following screenshot is a representation of audit logs that are accessible with the GCP console UI:

Figure 3.10 – Sample output audit logs through the GCP console

This audit log can be informative, providing different teams such as security, operations, and developers access to who (individual or automated service account) created builds. You can view a list of audited operations available for Cloud Build to gain more insight into how Cloud Build is being utilized at the following link: `https://cloud.google.com/build/docs/securing-builds/audit-logs#audited_operations`. Build provenance for auditing build information will be covered in *Chapter 6*, *Managing Environment Security*.

The `gcloud logging read` command can also be used to view build logs as well as other GCP services.

Kicking off a build is critical to the use of Cloud Build, but understanding which other operations can be performed or what kind of data is available is just as important.

Summary

Executing a build pipeline can be fairly straightforward once a valid configuration manifest is provided along with some settings for Cloud Build. The managed service takes care of the behind-the-scenes configuration validation, resource provisioning, and execution of the build itself. Given that it's a managed service, information is made available to provide users the status through various means, whether visually through the GCP console UI or a standard set of APIs.

Stay tuned for the next chapter as we dig into the manifest configuration for Cloud Build.

Part 2: Deconstructing a Build

In this part of the book, we will dive into writing and configuring full-featured builds. You will understand the core building blocks that make up a build configuration, the various ways these builds can be executed, and how to design these builds to run with security best practices in mind.

This part comprises the following chapters:

- *Chapter 4, Build Configuration and Schema*
- *Chapter 5, Triggering Builds*
- *Chapter 6, Managing Environment Security*

4

Build Configuration and Schema

We have discussed the versatility of the builds you can run in Cloud Build. As a user, you are able to define your own tasks that make up each build step in your build.

The way that you accomplish this is by writing build definitions according to a declarative schema that is specific to Cloud Build. This schema enables you to define not only individual build steps but also their sequence and relationship to one another, along with other specific build-wide configurations such as timeouts or permissions in the Cloud Build execution environment.

In this chapter, we will review the following topics:

- Defining the minimum configuration for build steps
- Adjusting the default configuration for the build steps
- Defining the relationships between individual build steps
- Configuring build-wide specifications

Defining the minimum configuration for build steps

In Cloud Build, builds are the primary resource you will deal with as a user. Typically, builds are defined via a configuration file written in YAML according to Cloud Build's schema, though JSON is also supported. There are also minor exceptions for creating a build without providing a configuration file, which we will review in *Chapter 9, Automating Serverless with Cloud Build*.

Let's review what these configuration files look like, starting with a build we are already familiar with to some degree.

Setting up your environment

In the previous chapter, we reviewed a build using a simplified version of the following `cloudbuild.yaml` config file, which can be found at `https://github.com/GoogleCloudPlatform/cloud-build-samples/tree/main/basic-config`.

This build configuration file defines three individual build steps that are carried out sequentially:

- Build a Docker container image.
- Push the Docker container image to **Artifact Registry** (`https://cloud.google.com/artifact-registry`).
- Create a **Compute Engine** (`https://cloud.google.com/compute`) VM and deploy the previously built container image to the newly created VM.

Additionally, we can look at a visual representation of the build here:

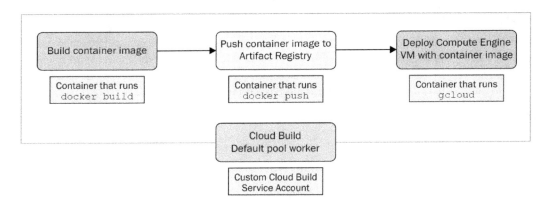

Figure 4.1 – A simple build that builds a container, pushes it to a registry, and runs it on a new VM

The contents in Google Cloud Platform's GitHub repositories might change over time, so let's create our own copy of the files needed to run the build exercises in this chapter. Let's utilize **Cloud Shell** (`shell.cloud.google.com`), the ephemeral Linux workstation provided by Google Cloud.

It is highly recommended that you run these exercises in a sandbox or test Google Cloud project, as you will have full control not only to run the exercises but also to easily clean up any of the resources created.

Once you have a Terminal session open in Cloud Shell, set the environment variables that we'll pass to our build configuration file:

```
$ PROJECT_ID=$(gcloud config get project)
```

```
$ SERVICE_ACCOUNT=packt-cloudbuild-chp4-sa
```

Next, create the working directory in which we will store the files for this chapter's build exercises:

```
$ mkdir packt && cd packt && \
mkdir cloudbuild && cd cloudbuild && \
mkdir chapter4 && cd chapter4 && \
mkdir first-build && cd first-build
```

Create the first build configuration file from the aforementioned build, which will build a container image, push it to Artifact Registry, and create a VM to run the container image:

```
$ cat > cloudbuild.yaml <<EOF
steps:
  - name: 'gcr.io/cloud-builders/docker'
    args: ['build', '-t',
           'us-central1-docker.pkg.dev/${PROJECT_ID}/my-docker-
repo/myimage',
           '.']

  - name: 'gcr.io/cloud-builders/docker'
    args: ['push',
           'us-central1-docker.pkg.dev/${PROJECT_ID}/my-docker-
repo/myimage']

  - name: 'gcr.io/google.com/cloudsdktool/cloud-sdk'
    entrypoint: 'gcloud'
    timeout: 240s
    args: ['compute', 'instances',
           'create-with-container', 'my-vm-name',
           '--container-image',
           'us-central1-docker.pkg.dev/${PROJECT_ID}/my-docker-
repo/myimage']
    env:
      - 'CLOUDSDK_COMPUTE_REGION=us-central1'
      - 'CLOUDSDK_COMPUTE_ZONE=us-central1-a'
serviceAccount: 'projects/${PROJECT_ID}/
serviceAccounts/${SERVICE_ACCOUNT}@${PROJECT_ID}.iam.
gserviceaccount.com'
```

```
images: [us-central1-docker.pkg.dev/${PROJECT_ID}/my-docker-
repo/myimage']
options:
  logging: CLOUD_LOGGING_ONLY


EOF
```

> **Note**
>
> When working with container images, it is best practice to specify the `images` field with the image you are creating. This enables functionality such as build provenance, which we will discuss in *Chapter 6, Managing Environment Security*.
>
> For simple builds that are okay with pushing the container image at the end of the build's run, you can omit a distinct build step to build this image and just specify the `images` field instead. Doing so will prompt Cloud Build to push the image in the field at the end of the build's run.
>
> However, in this example, because we are creating a VM that runs our newly built container image, we still maintain a distinct build step.

Create the Dockerfile for the container image we will build in this build:

```
$ cat > Dockerfile <<EOF
FROM alpine
CMD ["echo", "Hello World!"]
EOF
```

Earlier in the book, we briefly noted that builds utilize a **Service Account** from **Identity and Access Management (IAM)** to perform certain actions that access Google Cloud APIs. For this chapter, we will create a temporary service account:

```
$ gcloud iam service-accounts create packt-cloudbuild-chp4-sa \
    --description="Temporary SA for chp 4 exercises" \
    --display-name="packt-cloudbuild-chp4-sa"
```

Now, let's give the service account permissions that the builds will require in order to execute during this chapter's exercises:

```
$ gcloud projects add-iam-policy-binding ${PROJECT_ID} \
  --member="serviceAccount:packt-cloudbuild-chp4-sa@${PROJECT_
ID}.iam.gserviceaccount.com" \
  --role="roles/iam.serviceAccountUser" \
  --role="roles/cloudasset.owner" \
  --role="roles/storage.admin" \
  --role="roles/logging.logWriter" \
  --role="roles/artifactregistry.admin" \
  --role="roles/compute.admin
```

We now have the files and permissions necessary to work with in this chapter. Now, let's create a repository in Artifact Registry for the container images that will be used in this chapter:

```
$ gcloud artifacts repositories create my-docker-repo \
  --repository-format=Docker \
  --location us-central1
```

Now, you can kick off our first build exercise:

```
$ gcloud builds submit . --region=us-central1
```

There will be opportunities to tweak the build file based on the concepts we review and rerun the build. However, because running `gcloud` to create a VM twice will result in a build failure since the VM already exists, run the following command before resubmitting this build to Cloud Build:

```
$ gcloud compute instances delete my-vm-name --zone=us-
central1-a
```

Once you are done with this example, you can use the following command to delete the service account used:

```
$ gcloud iam service-accounts deletepackt-cloudbuild-chp4-sa
@${PROJECT_ID}.iam.gserviceaccount.com
```

Now, with the setup done, let's dive into the build configuration file. Each build step in this build runs in its own container, and all the build steps for a given build execute on the same Cloud Build worker.

Your builds might be more complex and contain a greater number of build steps than the provided example, but the way in which the build steps are defined remains consistent with this example. So, let's break the build down bit by bit.

Defining your build step container image

Let's start with the top-level field, `steps`. This field encompasses all the build steps that will make up a build. All three build steps in the example are defined by stanzas underneath the `steps` field.

> **Note**
> At the time of writing, the maximum number of build steps you can have in a single build is 100 build steps. You can find the most up-to-date limitations at `https://cloud.google.com/build/quotas#resource_limits`.

In any of these stanzas, we can see the fields that a user must define to specify a single build step. At a bare minimum, we need to define the following two fields:

- `name`

- `args`

Here, `name` is a field that references the specific image path to your builder image that will run your build step. When you specify your container image, your Cloud Build worker will pull this container image and run an instance of it using `docker run`.

In the preceding example, we can see that the first and second steps both utilize the `gcr.io/cloud-builders/docker` image and that the third step utilizes the `gcr.io/google.com/cloudsdktool/cloud-sdk` image.

Note that to run this build in the prior chapter, you actually did not have to build these images yourself. Rather, these images are a part of the pre-built container images provided and hosted by Google that contain common toolings such as `docker` or `gcloud`. These Google-provided publicly available images are called **cloud builders** (`https://cloud.google.com/build/docs/cloud-builders`).

At the time of writing, cloud builders provide pre-built images for the following tools:

- General purpose builders, such as `wget`, `curl`, and `git`

- Language-specific builders, such as `dotnet`, `go`, `mvn`, `javac`, `npm`, and `yarn`

- Google Cloud-specific builders, such as `gcloud`, `gsutil`, `gcs-fetcher`, and `gke-deploy`

- Container- and Kubernetes-specific builders, such as `docker` and `kubectl`

Because these cloud builders are made available publicly by Google, you can actually navigate to the image path and see all of the versions of the image that Google has published. For example, navigate to `gcr.io/cloud-builders/docker` in your browser:

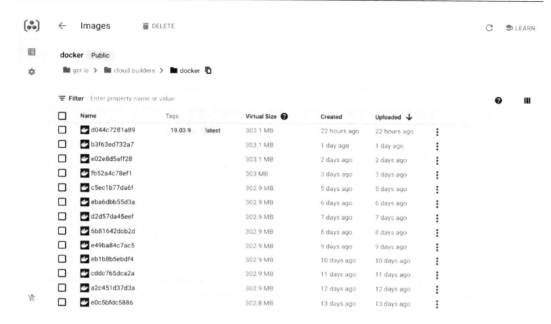

Figure 4.2 – Google-published container images for gcr.io/cloud-builders/docker

If you would like to customize these container images, you can view the provided source code (`https://github.com/GoogleCloudPlatform/cloud-builders`) and build your own variations of these images. For each builder, you can find a `Dockerfile` defining the container image for the cloud builder, along with a `cloudbuild.yaml` file to build your own version of the cloud builder.

If your use case for an individual build step is not covered by one of the provided cloud builder images, you can also turn to builders that are provided by the Cloud Build community (`https://github.com/GoogleCloudPlatform/cloud-builders-community`).

These are far greater in number and have varying cadences around the frequency of updates or required maintenance, as they are community owned. Some of the more commonly used community-contributed builders include the following:

- Container and Kubernetes builders, such as `helm`, `kustomize`, `ko`, and `pack`
- Infrastructure-as-code builders, such as `terraform`, `pulumi`, and `terragrunt`
- Mobile development builders, such as `firebase` and `flutter`

Similar to the official cloud builders, each of these community-contributed builders provides a `Dockerfile`, defining the builder container image, and a `cloudbuild.yaml` configuration file, to actually use Cloud Build to build the builder container image.

> **Note**
>
> If you are looking to either build your own variations of cloud builder images or build one of the community-provided builders, you will need to host these images inside your own repositories, such as Container Registry or Artifact Registry. You can even use third-party artifact stores such as JFrog's Artifactory.
>
> When doing so, it is a best practice that you are explicit about what version of a container you want to run for the build step using container image tags.
>
> This will enable you to ensure that you understand how a specific version of a container image will run your build. Additionally, it will ensure that your build does not break due to unforeseen changes in upstream container images.

Finally, you can utilize publicly available container images hosted at **Docker Hub** for the build steps. We will add a build step to our example using an image hosted at Docker Hub later in the *Adjusting the default configuration for the build steps* section.

Now that we've reviewed how you can use `name` to define the container images you'll use for a build step, let's talk about how you can more explicitly identify each build step using the `id` field.

In our existing `cloudbuild.yaml` file, we do not have any identifiers for each build step outside of the builder image in which it executes. This can be problematic for debugging builds or filtering through build logs if we reuse the same builder image multiple times in a build.

If you navigate to `console.cloud.google.com/cloud-build/builds` and visit the build you executed, by default, you'll see that each build step is identified by the builder image it uses:

Steps	Duration
✅ **Build Summary** 3 Steps	00:01:49
✅ 0: gcr.io/cloud-builders/docker build -t us-central1-docker.pkg.dev/agmsb-k8s/my-docker-repo/myima...	00:00:02
✅ 1: gcr.io/cloud-builders/docker push us-central1-docker.pkg.dev/agmsb-k8s/my-docker-repo/myimage	00:00:03
✅ 2: gcr.io/google.com/cloudsdktool/cloud-sdk gcloud compute instances create-with-container my-vm-name --contai...	00:01:36

Figure 4.3 – Build steps when the ID field is not defined per build step

Utilizing the `id` field in conjunction with the name field allows us to separately provide a human-friendly identifier for a build step. Running the same build with build step IDs would look like this:

```
steps:
  - name: 'gcr.io/cloud-builders/docker'
    id: 'Docker Build'
    args: ['build', '-t',
           'us-central1-docker.pkg.dev/${PROJECT_ID}/my-docker-
repo/myimage',
           '.']

  - name: 'gcr.io/cloud-builders/docker'
    id: 'Docker Push'
    args: ['push',
           'us-central1-docker.pkg.dev/${PROJECT_ID}/my-docker-
repo/myimage']

  - name: 'gcr.io/google.com/cloudsdktool/cloud-sdk'
    id: 'Deploy container to VM'
    entrypoint: 'gcloud'
    timeout: 240s
    args: ['compute', 'instances',
           'create-with-container', 'my-vm-name',
           '--container-image',
           'us-central1-docker.pkg.dev/${PROJECT_ID}/my-docker-
repo/myimage']
    env:
      - 'CLOUDSDK_COMPUTE_REGION=us-central1'
      - 'CLOUDSDK_COMPUTE_ZONE=us-central1-a'
```

If you would like to see this for yourself, feel free to tweak the `cloudbuild.yaml` example with different styles of defining IDs for each build step and rerun the build.

Now, you can utilize these build step IDs to more easily identify what each step is doing. Additionally, you can utilize them in a practical manner such as querying from **Cloud Logging** for logs from a specific build step across numerous runs of the build.

You can visit `console.cloud.google.com/logs` and enter a query, such as the following one, into the **Logs Explorer** to search for logs associated with a build step ID, filtering down to logs with the build step ID in their `textPayload`:

```
resource.type = "build"
textPayload =~ "Deploy container to VM"
```

Additionally, in the Cloud Build console, you can see that each build step is now referred to by its `id` field:

Steps	Duration
✅ **Build Summary** 3 Steps	00:01:32
✅ 0: Docker Build build -t us-central1-docker.pkg.dev/agmsb-k8s/my-docker-repo/myima...	00:00:01
✅ 1: Docker Push push us-central1-docker.pkg.dev/agmsb-k8s/my-docker-repo/myimage	00:00:03
✅ 2: Deploy container to VM gcloud compute instances create-with-container my-vm-name --contai...	00:01:20

Figure 4.4 – Build steps when the ID field is defined per build step

Using the `id` field will also be critical for defining the relationships between the build steps when they need to run in complex patterns, such as running multiple build steps in a specific order. We will review this later in the *Defining the relationships between individual build steps* section.

Now that we have reviewed how to provide the image in which your build step will run in the `name` field, and how to better identify your build steps using the `id` field, let's review the `args` field.

Defining your build step arguments

Here, `args` is the field in which you provide arguments to your cloud builder to use when carrying out tasks for a build step.

Note that `args` can be written in YAML using multiple styles:

- Flow style (`https://yaml.org/spec/1.2.2/#flow-sequences`)
- Block style (`https://yaml.org/spec/1.2.2/#block-sequences`)

In our example build configuration, the `args` field for each of the build steps is defined in flow style:

```
args: ['build',
       '-t',
       'us-central1-docker.pkg.dev/${PROJECT_ID}/my-docker-
repo/myimage',
       '.']
```

For smaller sets of arguments, flow style provides a simple and readable way to write `args`. You can also separate each item in flow style with a line break for additional readability.

However, for larger, more complex sets of arguments, utilizing block style might be easier to write. The same set of `args` fields from the previous example written in block style would look like the following:

```
args:
  -'build'
  -'-t'
  -'us-central1-docker.pkg.dev/${PROJECT_ID}/my-docker-repo/
myimage'
  -'.'
```

Feel free to tweak the `cloudbuild.yaml` example with different styles of defining `args` and rerun the build.

> **Note**
> The maximum number of `args` you can provide to a build step is 100, with the sum of characters in a single `args` field capped at 4,000.

Regardless of which way you write your arguments, there are two ways to use `args` to kick off having a builder carry out tasks:

- If the cloud builder provides an entrypoint, the values in `args` will be utilized as arguments for that entrypoint.
- If the cloud builder does not provide an entrypoint, the first value in `args` will be utilized as the cloud builder's entrypoint with the remaining values used as arguments.

Let's take a look at the builder image for our first and second build steps. If we view the source Dockerfile (https://github.com/GoogleCloudPlatform/cloud-builders/tree/master/docker), we can see that the container image has an ENTRYPOINT already defined. Because the content in Google Cloud Platform's GitHub repositories might change over time, we are also sharing the contents of the following Dockerfile:

```
FROM gcr.io/gcp-runtimes/ubuntu_20_0_4
ARG DOCKER_VERSION=5:19.03.9~3-0~ubuntu-focal
RUN apt-get -y update && \
  apt-get -y install \
  apt-transport-https \
  ca-certificates \
  curl \
  make \
  software-properties-common && \
  curl -fsSL https://download.docker.com/linux/ubuntu/gpg |
apt-key add - && \
  apt-key fingerprint 0EBFCD88 && \
  add-apt-repository \
  "deb [arch=amd64] https://download.docker.com/linux/ubuntu \
  $(lsb_release -cs) \
  edge" && \
  apt-get -y update && \
  apt-get -y install docker-ce=${DOCKER_VERSION} docker-ce-
cli=${DOCKER_VERSION}
ENTRYPOINT ["/usr/bin/docker"]
```

In the last line of the Dockerfile, we see ENTRYPOINT defined as ["/usr/bin/docker"], meaning the cloud builder will execute the Docker binary. Because this is the case, our first and second build steps provide arguments to instruct Docker to build a container image and push it to Artifact Registry.

However, this is not the case for the third build step in which we utilize the gcr.io/google.com/cloudsdktool/cloud-sdk builder. In this build step, we actually define the entrypoint using the entrypoint field in our cloudbuild.yaml. This is because the Dockerfile used to build the builder image does not define an entrypoint:

```
- name: 'gcr.io/google.com/cloudsdktool/cloud-sdk'
  entrypoint: 'gcloud'
```

You can use the entrypoint field in your configuration file for two main reasons:

- Explicitly defining the entrypoint if the builder container image does not specify it
- Overriding the ENTRYPOINT defined in the builder container image

One of the most common use cases for defining an entrypoint in your configuration file is being able to define a shell as the entrypoint and then pass arbitrary commands to the shell.

You could run the following as an example of this, which uses block style YAML and passes in `'-c'` as the first argument and then pipes a multiline string to run the rest of the arbitrary commands.

This example takes the public Docker Hub image for Python 3.9 and runs it as a build step, overriding the default python3 CMD to change the entrypoint to bash, while running a few commands in the shell:

```
$ cd .. && mkdir arbitrary-commands && cd arbitrary-commands
$ cat > cloudbuild.yaml <<EOF
steps:
  - name: 'python:3.9'
    id: 'Arbitrary commands in bash'
    entrypoint: 'bash'
    args:
      - '-c'
      - |
        pwd
        ls
        whoami
EOF
$ gcloud builds submit . --region=us-central1
```

Your output in the build logs should match the arbitrary commands defined in `args`:

```
/workspace
cloudbuild.yaml
root
```

This adds a layer of flexibility and configurability when running build steps have access to a shell.

Now that we have reviewed `name`, `id`, `args`, and `entrypoint` as the core functionality for how our build steps run, let's dig into additional ways to configure these build steps.

Adjusting the default configuration for the build steps

In addition to defining the container image and arguments for a build step, there are other configurations you might need to customize in order to carry out the tasks in that step in a specific manner.

These might be variables you want to set in the execution environment, making specific secrets available to the build step, or writing to a specific directory on the worker carrying out your build step.

The build step stanza has additional fields that cover cases such as the following:

- `env`
- `secretEnv`
- `timeout`
- `dir`

Let's start with the `env` field – this provides a way to set environment variables for the cloud builder in which your tasks execute. This provides a way for you to provide multiple environment variables that you might want to pass into the container that is running your tasks in a build step.

> **Note**
>
> Interestingly, `env` and `secretEnv` can be defined not only for an individual build step but also at a global level for all build steps in the build. While we will review configuring these globally in the *Configuring build-wide specifications* section, for any sensitive information, it is recommended that you configure these values at the individual build step.

In the third build step from our example, you can see that we define `env` to include two distinct variables:

- `CLOUDSDK_COMPUTE_REGION=us-central1`
- `CLOUDSDK_COMPUTE_ZONE=us-central1-a`

Given our entrypoint is running the `gcloud` CLI, defining these environment variables up front enables `gcloud` to bootstrap with these properties already defined. Specifically, with these already defined, the build step is able to create the VM that will run the container image in the us-central1-a zone.

This is representative of the typical use case for defining environment variables – preparing your environment for the build step to be able to carry out its tasks.

If we wanted to project a credential or a password into an environment variable, we would instead utilize `secretEnv`. We will review this in greater detail in *Chapter 8, Securing Software Delivery to GKE with Cloud Build*.

Not only do we see `env` defined in our build step stanza, but we also see `timeout` defined. This field is fairly straightforward; you define a period of time after which if the build step has not been completed, it will fail the entire build.

In the example we are working with, the build step that creates a VM has a timeout set to 240 seconds or 4 minutes. Timeouts are defined as a duration; the default unit is seconds if left unspecified. However, you can specify additional units via hours (h) and minutes (m).

Additionally, you have the ability to combine multiple units of time into a single timeout value. For example, `timeout: 2h10m45s` would be 2 hours, 10 minutes, and 45 seconds.

> **Note**
>
> At the time of publication, the maximum timeout for a build is 24 hours.

Timeouts are useful at the individual build step level if you have known build steps that depend on calling unreliable external systems or APIs, or when you are concerned about minimizing costs given that builds are priced per vCPU build minute utilized – especially when utilizing larger workers.

Similar to `env` and `secretEnv`, `timeout` can be defined not only for an individual build step but also at a global level for the entire build, which we will review in the *Configuring build-wide specifications* section.

Now, let's move on to the `dir` field. When a build executes, by default, the current working directory is set to `/workspace`. This is backed by a Docker volume mounted at this path by default, with the same volume provided to each build step that runs:

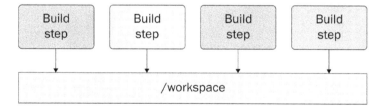

Figure 4.5 – Each build step will get access to the same volume mounted at /workspace

For many build steps, using `/workspace` as your working directory is sufficient. However, there are certain tools that need to run from a specific working directory, perhaps where certain configuration files are located that are passed to the tool.

There are a couple of examples of tools like this:

- Skaffold (`https://skaffold.dev/`) is a Kubernetes continuous development tool that is meant for Kubernetes users to run both in their local environment and further in their software delivery pipeline. To do this, users must execute the `skaffold` binary from the same directory in which there is a `skaffold.yaml` configuration file.

- Terraform (`https://www.terraform.io/`) is an infrastructure-as-code tool that is meant for users to utilize to create resources across multiple infrastructure providers using a consistent language. Usually, the `terraform` CLI should be executed from the working directory in which there are `.tf` configuration files, though it also has support for passing in `--chdir` as an argument that allows you to change the working directory.

For these scenarios, users can utilize the `dir` field to automatically set the working directory for a specific build step. There are two options in which you can use `dir`:

- Relative path
- Absolute path

Specifying `dir` as a relative path is the more common use case. Let's create a build that continues the work we did in our prior build and runs a couple of build steps in a new `cloudbuild.yaml` file:

```
$ cd .. && mkdir generate-validate-terraform && cd generate-
validate-terraform
$ PROJECT_ID=$(gcloud config get project)
$ ZONE=us-central1-a
$ cat > cloudbuild.yaml EOF<<
steps:
- name: 'gcr.io/google.com/cloudsdktool/cloud-sdk'
  id: 'Generate terraform for existing VMs'
  entrypoint: 'sh'
  args:
    - '-c'
    - |
      apt-get install google-cloud-sdk-config-connector
      gcloud beta resource-config bulk-export \
      --resource-format=terraform \
      --resource-types=ComputeInstance \
      --project=$PROJECT_ID \
      --path=tf
- name: 'hashicorp/terraform:1.0.0'
```

```
  id: 'Validate generated terraform'
  entrypoint: 'sh'
  args:
    - '-c'
    - |
      terraform init
      terraform validate
  dir: 'tf/projects/agmsb-k8s/ComputeInstances/$ZONE'
serviceAccount: 'projects/${PROJECT_ID}/
serviceAccounts/${SERVICE_ACCOUNT}'
EOF
$ gcloud builds submit . --region=us-central1
```

This build does two things:

- First, we are utilizing the functionality provided by the **Cloud Asset Inventory** API to export the resource we just created into a valid Terraform configuration file. Specifically, by not defining `dir`, we are utilizing `/workspace` as our working directory and creating a directory called `tf`.

- Second, we change the working directory to be where the previous build step exported our terraform configuration: `tf/projects/$PROJECT_ID/ComputeInstances/$ZONE`. This is a relative path, so we expect this directory to be under `workspace/`. Running the build step from this directory makes it easier to write cleaner `terraform` commands when passing them into a shell.

Additionally, this build provides examples of how we can apply the concepts we discussed in this chapter. Here are a couple of notes:

- We are using a Google-provided image for the first step: `gcr.io/google.com/cloudsdktool/cloud-sdk`. However, we also have to define `sh` as our entrypoint so that we can install a package necessary to run the example. This would be a prime opportunity for us to either pin our build to this specific version of the builder or build our own custom image that performs this installation outside of the build.

- We are using a public image in Hashicorp's registry in Docker Hub to utilize `terraform`. This is an opportunity to point out that Google does mirror the most popular images on Docker Hub to Container Registry. However, if pulling the image is too slow, we could also build our own builder using the community-provided terraform image.

Now that we have covered the other fields, we can configure to tweak our individual build steps, let's move on to see how we coordinate tasks across build steps in the same build.

Defining the relationships between individual build steps

Build steps rarely live in a vacuum and must understand their relationships to the other build steps in the build. A build step might need to write to files that a subsequent build step needs to access. Multiple build steps might need to run in parallel, wait for other build steps to complete before beginning their task, or even kick off a separate build.

Let's discuss two specific parameters that you could utilize to define these relationships between the build steps:

- volumes
- waitFor

volumes are the means by which data is persisted on the Cloud Build worker between build steps. We have already reviewed the default volume, workspace/, as the default working directory and the easiest place to persist files between build steps.

However, if your tooling needs to write files to a specific location, or expects files to be in a specific directory, you can also create your own volumes and mount them to persist files in your own custom locations.

An example of this might be when utilizing SSH keys in your build. Tooling, such as git, might expect SSH keys to be in /.ssh, which is under the home directory of the user that the container uses to run.

Specifying your own bespoke volumes to persist data between build steps would look like this:

```
steps:
- name: 'gcr.io/cloud-builders/git'
  secretEnv: ['SSH_KEY']
  entrypoint: 'bash'
  args:
  - -c
  - |
    echo "$$SSH_KEY" >> /root/.ssh/id_rsa
    chmod 400 /root/.ssh/id_rsa
    cp known_hosts.github /root/.ssh/known_hosts
  volumes:
  - name: 'ssh'
    path: /root/.ssh

# Clone the repository
- name: 'gcr.io/cloud-builders/git'
```

```
args:
- clone
- --recurse-submodules
- git@github.com:GIT_USERNAME/GIT_REPOSITORY
volumes:
- name: 'ssh'
  path: /root/.ssh
```

This means that you could use the `ssh` volume or the workspace volume to persist data between these two build steps. You can find a fully built-out tutorial for this example at `https://cloud.google.com/build/docs/access-github-from-build`.

Data might not be the only thing you want to orchestrate across build steps; you might want to have greater controls to orchestrate their execution, too.

This is where the `waitFor` field comes in handy. This field allows you to alter the default sequential execution of the build steps.

The `waitFor` field accepts two kinds of values:

- `-`: This means the build step will run immediately upon build execution.
- `id`: This means that the build step will wait for the build step with the specified `id` field to complete before executing. You can also specify multiple `id` fields if you want a build step to wait for more than one build step to complete.

This leads us to be able to make a more graph-like ordering of steps, where we can execute multiple build steps in a specific order, with the functionality to implement blocking at various stages in our build.

Here, we have a simple fan-in, fan-out pattern using the `waitFor` field for each build step:

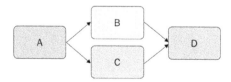

Figure 4.6 – Using waitFor to fan out to concurrent build steps and then fan in

Now we can execute this build using simple `bash` builder images to demonstrate how Cloud Build handles the `waitFor` fields in each build step.

Keep in mind that while these steps can occur in parallel, it is not guaranteed. Rather, it is the *ordering* of build step execution that is respected:

```
$ cd .. && mkdir fan-out-fan-in && cd fan-out-fan-in
$ cat > cloudbuild.yaml <<EOF
steps:

- name: 'bash'
  id: A
  args:
  - 'sleep'
  - '10'
  waitFor: ['-']

- name: 'bash'
  id: B
  args:
  - 'sleep'
  - '20'
  waitFor: ['A']
- name: 'bash'
  id: C
  args:
  - 'sleep'
  - '20'
  waitFor: ['A']
- name: 'bash'
  id: D
  args:
  - 'sleep'
  - '30'
  waitFor: ['B', 'C']
EOF

$ gcloud builds submit . --region=us-central1
```

In this example, we have step A execute immediately, while both B and C wait for A to complete its execution. Upon completion, both start in parallel, while step D waits for both B and C to complete before execution.

Additionally, we can use this same functionality to execute preparation-like steps, such as grabbing files from multiple locations that a later build step might need to carry out a task. Removing blocking functionality by stating that A-1, A-2, and A-3 are allowed to run at the same time could potentially speed up the build and can also help cut costs by reducing the overall build minutes. However, while it is possible, there is no guarantee that they will all run in parallel:

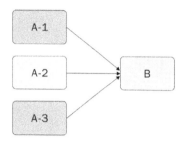

Figure 4.7 – Using waitFor to run multiple build steps concurrently and immediately

Having now discussed sharing data between build steps and orchestrating the ordering of build steps, let's now review global configurations that are not specific to any one build step.

Configuring build-wide specifications

Finally, there are sets of configurations that you might use to configure global settings that apply across all build steps. We have already discussed some of them prior to this chapter:

- `timeout`
- `env`
- `secretEnv`

The preceding fields behave the same as their counterparts defined at the individual build step level; the core difference is that defining them globally now makes them available to all build steps in the build.

However, there is an additional global field that you must specify when you are using `secretEnv`, regardless of whether it is at the individual build step level or the global level:

- `availableSecrets`

It is in this field where you declare the `availableSecrets` you would like to use at some point in your build, as well as where you are pulling them from. It is recommended that you utilize **Secrets Manager** to store sensitive credentials that you will be using in your builds; we will review this, in detail, in *Chapter 8, Securing Software Delivery to GKE with Cloud Build*.

You can find the full list of configuration fields for a build in Cloud Build at `https://cloud.google.com/build/docs/build-config-file-schema`. The full schema at publication time is outlined next:

```
steps:
- name: string
  args: [string, string, ...]
  env: [string, string, ...]
  dir: string
  id: string
  waitFor: [string, string, ...]
  entrypoint: string
  secretEnv: string
  volumes: object(Volume)
  timeout: string (Duration format)
- name: string
  ...
- name: string
  ...
timeout: string (Duration format)
queueTtl: string (Duration format)
logsBucket: string
options:
  env: [string, string, ...]
  secretEnv: string
  volumes: object(Volume)
  sourceProvenanceHash: enum(HashType)
  machineType: enum(MachineType)
  diskSizeGb: string (int64 format)
  substitutionOption: enum(SubstitutionOption)
  dynamicSubstitutions: boolean
  logStreamingOption: enum(LogStreamingOption)
  logging: enum(LoggingMode)
```

```
  pool: object(PoolOption)
  requestedVerifyOption: enum(RequestedVerifyOption)
substitutions: map (key: string, value: string)
tags: [string, string, ...]
serviceAccount: string
secrets: object(Secret)
availableSecrets: object(Secrets)
artifacts: object (Artifacts)
images:
- [string, string, ...]
```

While we did not review every single field in this schema, we will continue to cover the remainder of configurations in practical applications in *Chapters 7, 8, 9*, and *10*.

Summary

The Cloud Build configuration schema provides an intuitive, flexible, and integrated method for defining build steps and builds. Whether you are looking to merely run simple and common tasks covered by cloud builders, or to build and run your own custom build steps for complex pipelines, there is much configurability in Cloud Build's configuration.

Now that we can write fairly complex builds using the schema, let's dive into how we can trigger the execution of these builds based on certain events, in *Chapter 5, Triggering Builds*.

5

Triggering Builds

One of the key parts of automation is establishing how a pipeline is to be initiated. This is configured by setting up a trigger for each pipeline. In *Chapter 3, Getting Started – What Build Information Is Available to Me?*, we covered how to initiate a build manually with the CLI using `gcloud`. Cloud Build offers a variety of ways to trigger builds, whether integrated with services offered by **Google Cloud Platform** (**GCP**) or external mechanisms such as a **webhook**. In this chapter, we will see the various trigger options offered by Cloud Build as we traverse through the following topics:

- The anatomy of a trigger
- Integrations with source code management (SCM) systems
- Defining your own triggers

Technical requirements

- Software build concepts and tools
- Git repositories

The anatomy of a trigger

A Cloud Build trigger requires a few pieces of information at a minimum in order to be properly configured. A trigger requires the following data points:

- Name
- Region
- Event triggers (defaults to Push to a branch):
  - Push to a branch
  - Push new tag
  - Pull request

- Other mechanisms
  - Manual invocation – triggered via the GCP console or automated means (CLI, REST, or SDK)
  - Google Cloud Pub/Sub message
  - Webhook event – which is an HTTP post to a designated URL
- Source repository
- Source branch/tag:
  - Regular expressions can be used to determine which branch/tag names trigger an event.
- Cloud Build configuration file (defaults to `cloudbuild.yaml`):
  - Dockerfile – if you want to build a container
  - Buildpacks – which can detect the source code language and build a container based on the defined specification

Other fields can also play a vital role in helping to customize the trigger:

- Variables – used for substitution in the build configuration. This can override variables pre-defined in the Cloud Build configuration file (`yaml` or `json`).
- Approval – will determine whether an approval is required before the build can begin. This can be done in the GCP UI console or via automated means.
- Service account – allows the worker to use a specified service account that contains restricted or specific privileges for the build.

The repository in a trigger is critical because it needs the source of the resources used in the build. In the previous chapter, we were able to manually submit a build via the CLI without a repository because we referenced the local path. The CLI packaged up the path and its contents for us to be used as part of the build.

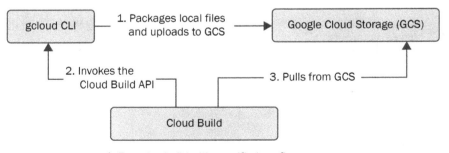

4. Executes build with specified config

Figure 5.1 – Example flow when a build is manually triggered via
the CLI and the source is in the local filesystem

When a trigger is preconfigured with an SCM repository, the build will fetch the code from the integrated repository for that specific branch/tag and commit if available.

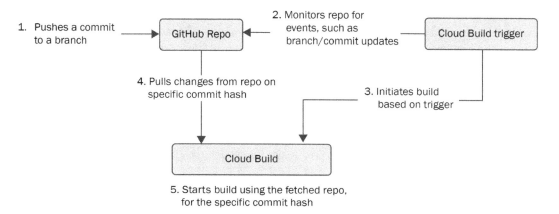

Figure 5.2 – The flow when a build fetches the repository from the configured SCM

The commands and output in this chapter are for illustrative purposes to describe capability as well as analysis of the output where applicable. Some of these commands may require prerequisites configured in order to execute properly.

A trigger can be created for GitHub using the CLI in the following example:

```
$ gcloud builds triggers create github \
    --name=${TRIGGER_NAME} \
    --repo-name=${REPO_NAME} \
    --repo-owner=${REPO_OWNER} \
    --branch-pattern=${REGEX} \
    --build-config="cloudbuild.yaml"
```

The CLI, REST, and SDK can be used to create, describe, update and delete triggers. In the preceding example, it is expected that the Cloud Build GitHub app has authorized access to the specified ${REPO_NAME}.

We have just gone through the flexibility of a trigger that allows for varying levels of customization. Next, we will dig deeper into the built-in integrations with different source code management platforms.

Integrations with source code management platforms

The source repository is a critical component of a trigger because it's dependent on the resources in the repository and a snapshot of the files so that builds can reference a particular set of source files. Cloud Build supports a growing list of SCM platforms. Following is a list of currently supported SCM platforms and how they are integrated with Cloud Build (that is, trigger types and how repositories can be accessed).

SCM Platform	Trigger type	Access
Cloud Source Repositories	Native	Service account permissions
GitHub	Native	GitHub auth and Cloud Build GitHub app
GitHub Enterprise	Native	GitHub app and enterprise config
GitLab	Webhook	SSH key in Secret Manager
GitLab Enterprise	Native	GitLab auth and enterprise config
Bitbucket Cloud	Native	SSH key in Secret Manager
Bitbucket Server	Native	Tokens in Secret Manager and enterprise config
Bitbucket Data Center	Native	Tokens in Secret Manager and enterprise config

Table 5.1 – SCM platform trigger type and access configurations

Each of these platforms, as noted, may have a native trigger or webhook trigger configuration. The authentication and authorization of repositories may also depend on the platform. When there are secret resources to be managed, native **Google Secret Manager** integration is available to help manage secrets. For **GitHub**, via the GitHub app, it is also possible to update the build status on the commit in the repository.

Once a trigger is configured to listen on a specific regex of the repository branch or tag, a matching commit will initiate a build.

The following exports an existing trigger:

```
$ gcloud builds triggers export \
    packt-cloudbuild-trigger \
    --destination export.yaml
```

Here's the output of the generated `export.yaml` configuration:

```
autodetect: true
createTime: '2022-01-04T20:26:51.731858991Z'
description: Invokes a build every time code is pushed to any
branch
```

```
github:
  name: packt-cloudbuild
  owner: kenthua
  push:
    branch: ^main$
id: 2aad99c0-2e3c-4a43-9b24-cbdaa12cbaf8
name: packt-cloudbuild-trigger
tags:
- cloudbuild-sample-branch-trigger
```

The code is committed to the branch that the trigger is listening to:

$ git push origin main

Example output of the push follows. Note the commit hash of 5023f77:

```
Enumerating objects: 5, done.
Counting objects: 100% (5/5), done.
Delta compression using up to 4 threads
Compressing objects: 100% (3/3), done.
Writing objects: 100% (3/3), 411 bytes | 411.00 KiB/s, done.
Total 3 (delta 1), reused 0 (delta 0)
remote: Resolving deltas: 100% (1/1), completed with 1 local
object.
To github.com:kenthua/packt-cloudbuild.git
   17ab064..5023f77  main -> main
```

Given the preceding example trigger, once code is pushed to the main branch, a build will trigger and pull the source. Listing the build history using the CLI outputs the following:

$ gcloud builds list

The following is a truncated list of the build in progress. You are able to query the status and track the build as needed:

```
ID: a50ec9b7-84a1-467e-b5ca-d89af472f783
CREATE_TIME: 2022-02-10T01:48:44+00:00
DURATION: 2M4S
SOURCE: -
```

```
IMAGES: -
STATUS: WORKING
```

The following command retrieves the logs of the specified build:

```
$ gcloud builds log a50ec9b7-84a1-467e-b5ca-d89af472f783
```

The following is a truncated log output of the build:

```
-------------------------------------------- REMOTE BUILD
OUTPUT ----------------------------------------------
starting build "a50ec9b7-84a1-467e-b5ca-d89af472f783"


FETCHSOURCE
hint: Using 'master' as the name for the initial branch. This
default branch name
hint: is subject to change. To configure the initial branch
name to use in all
hint: of your new repositories, which will suppress this
warning, call:
hint:
hint:    git config --global init.defaultBranch <name>
hint:
hint: Names commonly chosen instead of 'master' are 'main',
'trunk' and
hint: 'development'. The just-created branch can be renamed via
this command:
hint:
hint:    git branch -m <name>
Initialized empty Git repository in /workspace/.git/
From https://github.com/kenthua/packt-cloudbuild
 * branch            5023f77a3913ac3c97ec0bacac7257fa29b60a5a ->
FETCH_HEAD
HEAD is now at 5023f77 update cloud build manifest;
BUILD
Starting Step #0
```

Note that it fetches the commit hash 5023f77, which was pushed to the SCM repository. Cloud Build itself doesn't have the build resources as it was triggered by the SCM event, but the build still needs to fetch the commit from the repository to proceed with the build.

Defining your own triggers

Pre-built integrations are very helpful for getting started quickly. However, they are not the only way to integrate with Cloud Build. You can leverage other mechanisms such as a manual or webhook trigger in order to integrate with existing systems or platforms that currently are not yet pre-built in Cloud Build.

Webhook triggers

An example we will be running through is a webhook trigger and we will be using GitLab as an example. Native integration with GitLab offerings are on the way as noted earlier in the chapter.

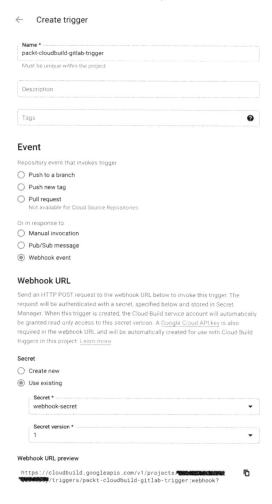

Figure 5.3 – An example of webhook trigger creation

Let's break down the webhook URL:

- Domain: cloudbuild.googleapis.com
- Google Project ID
- Trigger Name: packt-cloudbuild-gitlab-trigger
- Type: webhook
- Key: generated API key
- Secret: generated secret when specifying **New** or an existing secret such as a password

In the example URL noted previously, we have a standard domain, the context path consisting of the Google project ID, and a trigger name. The unique elements protecting the URL are the key and secret. The URL is expecting an HTTP POST method and optional data, which can be parsed to be used in the pipeline.

Advanced

Substitution variables

Substitutions allow re-use of a cloudbuild.yaml file with different variable values. Use bash string manipulation to combine variables and bindings to access arbitrary data in the JSON payload of the webhook. Learn more

Figure 5.4 – Substitution variables

Substitution variables can be used to allow for flexibility of common pipeline files with variables that can be passed in to make the pipeline more flexible. In this case, we are using a substitution variable to extract the GitLab commit ID to be used in the pipeline. As noted previously, in the case of GitLab, the response consists of repository metadata and most importantly the commit ID, which can be used in the pipeline to retrieve the specific commit (https://docs.gitlab.com/ee/user/project/integrations/webhook_events.html#push-events).

In a webhook configuration or when trying to access repositories that do not have pre-built integration, Cloud Build is not able to access the repository for a cloudbuild.yaml configuration. In this case, a trigger must have the pipeline configuration inline. This can be edited via the GCP console UI. If using the gcloud CLI, it can be added using an --inline-config switch, which references a cloudbuild.yaml configuration in the filesystem.

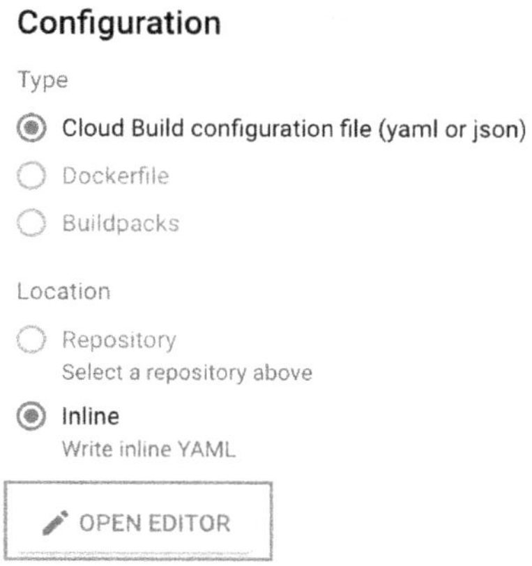

Configuration

Type

◉ Cloud Build configuration file (yaml or json)

○ Dockerfile

○ Buildpacks

Location

○ Repository
Select a repository above

◉ Inline
Write inline YAML

✏ OPEN EDITOR

Figure 5.5 – Build configuration

Selecting **OPEN EDITOR** in the preceding figure will open up an editor for specifying the pipeline configuration. The following screenshot is an example configuration snippet of the inline `cloudbuild.yaml` for the trigger when adding through the GCP console UI.

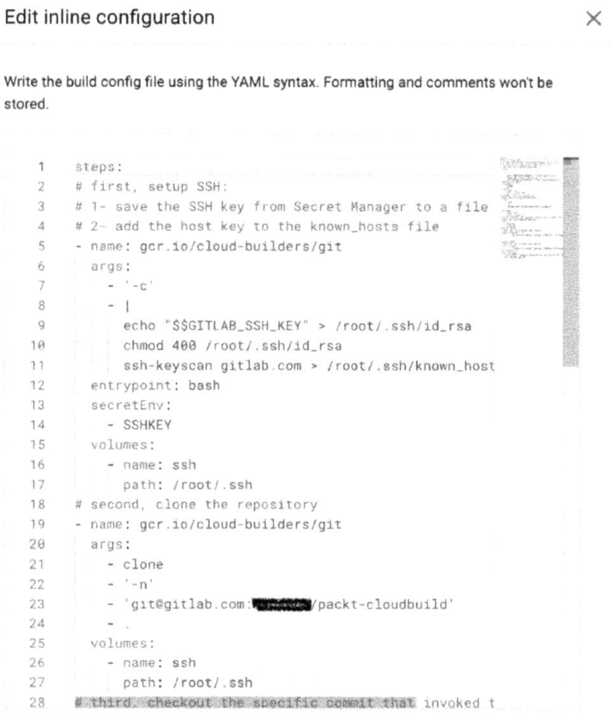

Edit inline configuration   ✕

Write the build config file using the YAML syntax. Formatting and comments won't be stored.

```
1   steps:
2   # first, setup SSH:
3   # 1- save the SSH key from Secret Manager to a file
4   # 2- add the host key to the known_hosts file
5   - name: gcr.io/cloud-builders/git
6     args:
7       - '-c'
8       - |
9         echo "$$GITLAB_SSH_KEY" > /root/.ssh/id_rsa
10        chmod 400 /root/.ssh/id_rsa
11        ssh-keyscan gitlab.com > /root/.ssh/known_host
12    entrypoint: bash
13    secretEnv:
14      - SSHKEY
15    volumes:
16      - name: ssh
17        path: /root/.ssh
18  # second, clone the repository
19  - name: gcr.io/cloud-builders/git
20    args:
21      - clone
22      - '-n'
23      - 'git@gitlab.com:████████/packt-cloudbuild'
24      - .
25    volumes:
26      - name: ssh
27        path: /root/.ssh
28  # third, checkout the specific commit that invoked t
```

Figure 5.6 – Inline build configuration editor

The preceding screenshot shows the pipeline configuration in the UI editor. The following code is a complete example of the `cloudbuild.yaml` configuration:

```yaml
steps:
[1]- name: gcr.io/cloud-builders/git
    args:
      - '-c'
      - |
        echo "$$GITLAB_SSH_KEY" > /root/.ssh/id_rsa
        chmod 400 /root/.ssh/id_rsa
        ssh-keyscan gitlab.com > /root/.ssh/known_hosts
    entrypoint: bash
    secretEnv:
      - GITLAB_SSH_KEY
    volumes:
      - name: ssh
        path: /root/.ssh
[2]- name: gcr.io/cloud-builders/git
    args:
      - clone
      - '-n'
      - 'git@gitlab.com:##redacted_repo_org##/packt-cloudbuild'
      - .
    volumes:
      - name: ssh
        path: /root/.ssh
[3]- name: gcr.io/cloud-builders/git
    args:
      - checkout
      - $_TO_SHA
  - name: gcr.io/cloud-builders/docker
    args:
      - build
      - '-t'
      - 'us-central1-docker.pkg.dev/${PROJECT_ID}/image-repo/
myimage'
      - .
```

```
    - name: gcr.io/cloud-builders/docker
      args:
        - push
        - 'us-central1-docker.pkg.dev/${PROJECT_ID}/image-repo/
myimage'
availableSecrets:
[4]  secretManager:
      - versionName: projects/##REDACTED-project_number/secrets/
GITLAB_SSH_KEY/versions/1
        env: GITLAB_SSH_KEY
```

The highlighted numbers have been added for reference. We will use them to walk through and describe each of the steps:

1. This step prepares a secure Git connection and stores the GitLab ssh key variable's GITLAB_SSH_KEY contents in a mounted volume.

2. Clone the specified private repository using the key stored in the mounted volume.

3. Retrieve the contents of the specific commit SHA by using the checkout command. The SHA is retrieved from the `$_TO_SHA` substitution variable specified in the trigger configuration. If you recall, the variable references the `${body.after}` value, which comes from the HTTP POST of the webhook URL triggered by GitLab. Cloud Build parses the JSON body content and makes it available to the pipeline.

4. Retrieve the GitLab ssh key that is stored in **Secret Manager**. The secret context path is specified to locate the specific secret. It is then associated to the **GITLAB_SSH_KEY** variable.

This sample cloudbuild.yaml configuration has a few steps to clone and check out the specific commit tag. Once the repository is accessible by Cloud Build, we can continue with the steps to complete the pipeline.

Manual triggers

A manual trigger can be invoked with multiple mechanisms to expand the types of custom integrations. An existing system tool with the appropriate credentials that is critical to the build pipeline can be integrated with a manual trigger using the gcloud CLI or the REST API.

The following is an example execution of a manual trigger using the gcloud command:

```
$ gcloud beta builds triggers run [trigger_name] \
--region global
```

Another mechanism that leverages the manual trigger is when build pipelines need to execute at a specific time or interval. Cloud Scheduler has direct integration with Cloud Build in order to execute a Cloud Build manual trigger at a specified time. From a Cloud Build manual trigger, you can run a trigger on a schedule, which can be created via the UI:

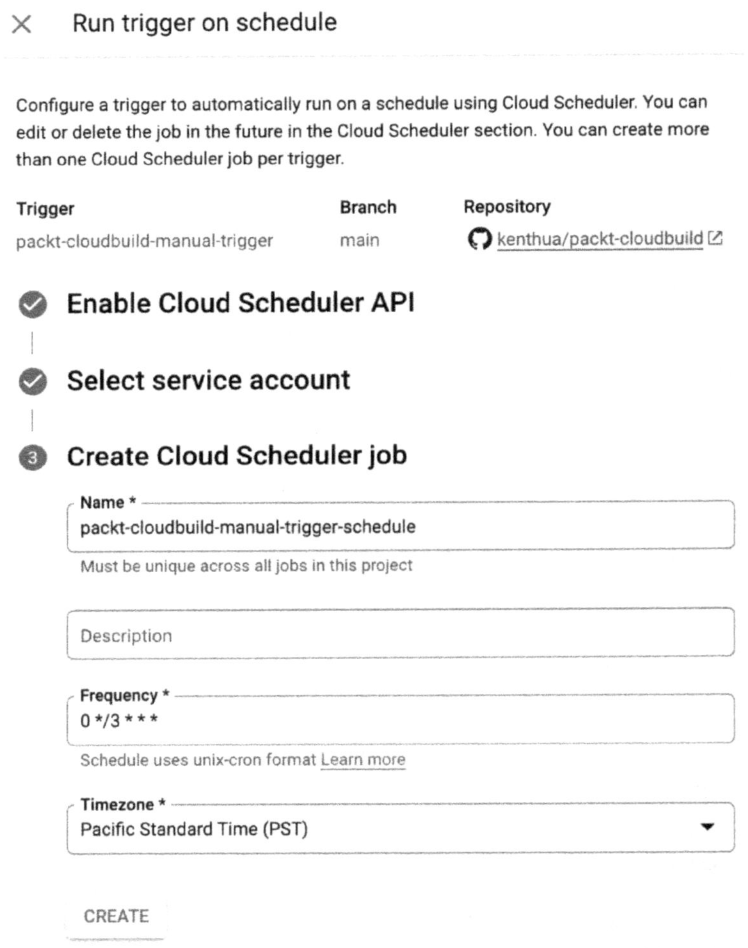

×   Run trigger on schedule

Configure a trigger to automatically run on a schedule using Cloud Scheduler. You can edit or delete the job in the future in the Cloud Scheduler section. You can create more than one Cloud Scheduler job per trigger.

Trigger	Branch	Repository
packt-cloudbuild-manual-trigger	main	○ kenthua/packt-cloudbuild ☑

✓ **Enable Cloud Scheduler API**

✓ **Select service account**

③ **Create Cloud Scheduler job**

Name *
```
packt-cloudbuild-manual-trigger-schedule
```
Must be unique across all jobs in this project

```
Description
```

Frequency *
```
0 */3 * * *
```
Schedule uses unix-cron format Learn more

Timezone *
```
Pacific Standard Time (PST)                                    ▼
```

CREATE

Figure 5.7 – Cloud Scheduler job creation

Going through the wizard, you select the service account that will be used by Cloud Scheduler to invoke the trigger. A default service account is created or one that is set up with the appropriate permissions can be specified.

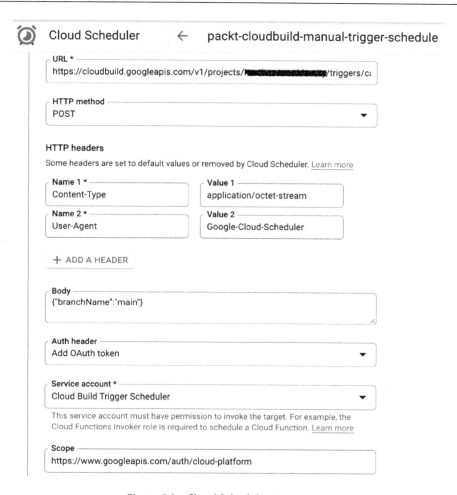

Figure 5.8 – Cloud Scheduler integration

The preceding screenshot is the **Cloud Scheduler** job that is generated from the Cloud Build manual trigger. Notice that Cloud Scheduler users the Cloud Build REST API to invoke the manual trigger.

Summary

Cloud Build provides multiple ways to integrate with existing system tools and SCM platforms to participate in automated build pipelines. Cloud Build triggers provide the ability to execute a build pipeline. This can be through SCM events such as committing to a branch, external factors such as being the smaller part of a larger automation pipeline of another system tool, or at a specified time.

In this chapter, we showed an example of storing an SSH key in Secret Manager to be utilized by the build pipeline. In the next chapter, we will go further into security best practices with Cloud Build.

6
Managing Environment Security

In a managed service, shared responsibility for securing the resources must be factored in. The provider is responsible for securing the infrastructure, managing services that run on top, and providing security-related capabilities. The consumer is responsible for who has access to the service and securing the data outside of the managed service.

In this chapter, we will cover the following topics of how Cloud Build provides security while also integrating with other security-related services:

- Defense in depth
- The principle of least privilege
- Accessing sensitive data and secrets
- Building metadata for container images
- Securing the network perimeter

Defense in depth

The concept of defense in depth for computing is used by information security to ensure that security constructs are put in place at each layer for protection. Cloud Build ensures that these security constructs are in place; some examples are noted in *Figure 6.1*:

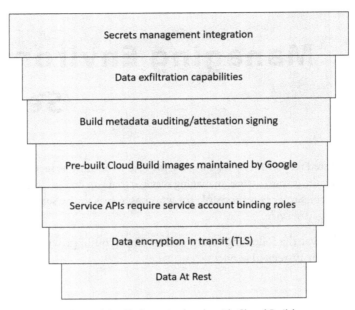

Figure 6.1 – Defense in depth with Cloud Build

While Cloud Build provides the capabilities mentioned here, some services must be enabled or leveraged in the pipeline. There are other security-related solutions available in the market that may not have direct integration with Cloud Build, but they can still be used for pipeline steps due to Cloud Build's support for custom container images in each step. For instance, you may be able to retrieve values from HashiCorp Vault if your container image has the correct libraries and tools.

There are also a few other examples to be aware of that may leave pipelines sharing sensitive information:

- Steps that output and log sensitive information
- Steps that use non-secure protocols to communicate with other systems
- Using unknown or vulnerable container images for steps within your pipeline
- Storing sensitive information in the Cloud Build working directory between steps

Defense in depth is a component of the overall security strategy; thus, the shared responsibility model also applies. Cloud Build, as noted before, provides many elements to secure your pipelines; however, certain steps and measures must be taken by the consumer of the service as well.

The principle of least privilege

The principle of least privilege is another security construct for protecting resources. The goal is to only provide the necessary access to resources to complete the job. If a pipeline does not need access to data in object storage, then there is no reason to grant access to the actor invoking the pipeline. For instance, organizations may be less restrictive about security permissions in the development phase, but more restrictive in the production phase. While this may make it easier to get things started, inconsistencies may cause trouble as teams progress to higher-level environments, causing unnecessary troubleshooting tasks. The cultural movement within organizations of shifting left with security from the onset may involve the concept of least privileged access to resources. This varies from organization to organization, as different regulatory bodies and industries may have different requirements.

Imagine if a bad actor gains access or someone in the organization unintentionally uses the service account. *Figure 6.2* depicts the scenario in visual form: they would be able to write to **Service A**, affecting the integrity of the data, while also being able to access or write data into **Service B** when it's not needed to complete a pipeline:

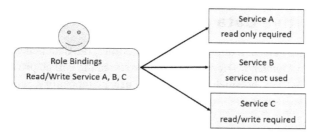

Figure 6.2 – Additional access granted when not needed

When leveraging the principle of least privilege, access is only granted to services that are necessary for a specific pipeline. *Figure 6.3* depicts that though a service exists, access is not granted because it is not needed in the pipeline. While this is in the context of a build pipeline, this can be applied throughout the organization for all services:

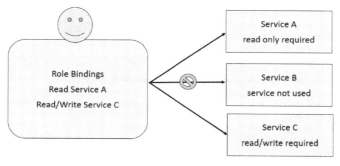

Figure 6.3 – Restricting access to a service to only what is needed

Cloud Build leverages the service accounts within **Google Cloud Platform** (GCP) to identify the actor that will be used in the pipeline to connect to other Google Cloud services. The service account can be assigned to each trigger, which allows for the same pipeline configuration, for instance, to be executed by different service accounts depending on the parameters that triggered the build. If you recall, in *Chapter 5, Triggering Builds*, Cloud Build offers regex patterns to determine how to trigger builds or different event invocation patterns. This allows us to use different service accounts on different Git branch, tag, and pull request patterns, webhooks, and manual invocations. These different types of triggering mechanisms were covered in *Chapter 5, Triggering Builds*.

Organizations can create service accounts with different role privileges and fine-grained permissions to define the minimal permissions required to successfully complete the pipeline. Leveraging unique service accounts also facilitates life cycle management, as well as provides administrators with the ability to monitor usage patterns within the defined service accounts.

Limiting access to resources through the principle of least privilege can reduce the exposure of an organization while also reducing the accidental manipulation of resources that was not intended.

Accessing sensitive data and secrets

Data is critical between the steps in a build pipeline; sometimes, the data being used in the pipeline may be sensitive as well. In *Chapter 5, Triggering Builds*, we stored the GitLab private SSH key in Secret Manager. The secret was used in the pipeline to clone the private repository. Sensitive data or secrets can be retrieved from various sources. Cloud Build has integrations with two GCP services for secrets:

- Secret Manager
- Cloud Key Management

It is important to protect the secret safely in a location that is not specified in the build configuration. Each of the respective services here also emits audit and data access logs that share when and which principal attempted to access a secret or key. Access to sensitive secrets using both of the aforementioned services is logged in Cloud Build for auditing purposes.

Secret Manager

The Secret Manager integration for Cloud Build is referenced by a stanza in `cloudbuild.yaml`. An example snippet from the previous chapter is shown here:

```
- name: gcr.io/cloud-builders/git
  args:
    - '-c'
    - |
      echo "$$GITLAB_SSH_KEY" > /root/.ssh/id_rsa
```

```
        chmod 400 /root/.ssh/id_rsa
        ssh-keyscan gitlab.com > /root/.ssh/known_hosts
    entrypoint: bash
    secretEnv:
    - GITLAB_SSH_KEY
...
availableSecrets:
  secretManager:
    - versionName: projects/282614098327/secrets/GITLAB_SSH_
KEY/versions/1
      env: GITLAB_SSH_KEY
```

The secretManager key is used to identify that the Secret Manager service is to be utilized. versionName is the unique identifier to the secret and version that is to be retrieved. The value of the secret is then stored in the environment variable, GITLAB_SSH_KEY, which can be used in the build steps of the pipeline. In this build step, notice how the variable named GITLAB_SSH_KEY is associated with the secretEnv key, so it can be leveraged in the args build argument.

Specifying availableSecrets in Cloud Build does not automatically grant access to the secret that is specified. The roles/secretmanager.secretAccessor role binding for the specific secret must be specified – in this case, referencing the service account to be used by the Cloud Build trigger.

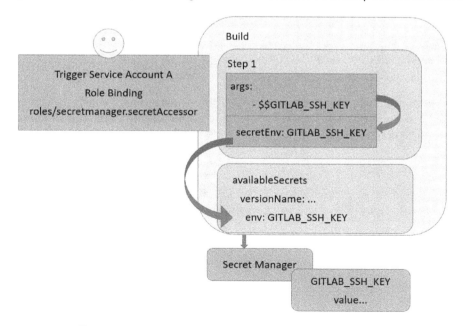

Figure 6.4 – How a secret is accessed by a build and build step

Figure 6.4 depicts how a build pipeline is associated with a specific service account with the appropriate role binding. The build can then retrieve the secret from Secret Manager and make it available within the build step using `secretEnv`.

Cloud Key Management

Cloud Build, as noted, can also use another GCP service, Cloud Key Management, to access sensitive data. The following consists of the example syntax used to access an encryption key from Cloud Key Management to decrypt an encoded secret:

```
availableSecrets:
  inline:
  - kmsKeyName: projects/**PROJECT_ID_redacted**/locations/
global/keyRings/secret_key_ring/cryptoKeys/secret_key_name
    envMap:
      MY_SECRET: '**ENCRYPTED_SECRET_encoded**'
```

Rather than using the `secretManager` reference, as in the previous example, the `inline` key is used to identify that the Cloud Key Management service is to be used. Notice some of the differences from the example snippet here in comparison to the Secret Manager scenario. In the Cloud Key Management scenario, `kmsKeyName` references the unique identifier to the encryption key. Cloud Build will use this key to decrypt the text specified as the value of `MY_SECRET`. The value of the secret is stored in `cloudbuild.yaml`, but in its encrypted form.

Just as with Secret Manager, Cloud Build also requires permission to the key through the Cloud Key Management service. In the case of Cloud Key Management, we require the following role for the Cloud Build trigger service account, `roles/cloudkms.cryptoKeyDecrypter`.

Cloud Build provides integration with multiple GCP cloud services for access-sensitive secrets and encrypted data. Sensitive data should not be stored in the build configuration or **source code management (SCM)** repositories.

Note

While we have provided instructions for Cloud Key Management, it is recommended to use Secret Manager to store and retrieve sensitive information for Cloud Build.

Build metadata for container images

Cloud Build generates metadata for container images that can be used to identify build details, build steps, attestations, or repository sources. SCM repository commit hashes and build ID hashes can be used to track the repository source and steps for each build. Cloud Build can generate and sign attestations at build time to allow organizations to enforce that only builds built-by-cloud-build can be deployed at runtime. We will cover each concept in the following sections.

Provenance

Cloud Build, in conjunction with Artifact Registry, can associate additional metadata about container images to validate the build details. For the most part, those that have access to the SCM and Cloud Build can use the metadata and logs to validate which SCM repository commit was used to trigger a specific build. As an added layer of auditing available to organizations, build provenance allows for organizations to validate that the metadata identifying the source and build steps of a container image has not been tampered with or manipulated. This metadata is stored in the format defined by the **Supply-chain Levels for Software Artifacts** (**SLSA**) framework (https://slsa.dev/provenance). Cloud Build generates this metadata and signs it so that organizations can validate the contents if desired. At the time of writing, Cloud Build can achieve an SLSA level of 2 (L2), which is defined as "tamper resistance of the build service" (https://slsa.dev/spec/v0.1/levels#summary-of-levels) and utilizes Cloud Build triggers with SCM repositories that have built-in integrations.

> **Note**
> For Cloud Build to generate the metadata, the following services must be enabled: Artifact Registry and Container Analysis. The cloudbuild.yaml configuration must also include the images field. For private pools to enable build provenance, the requestedVerifyOption option must be set to VERIFIED.

Artifact Registry provides the API to display the provenance metadata generated by Cloud Build with the -show-provenance switch. See an example command here:

```
$ gcloud artifacts docker images describe \
us-central1-docker.pkg.dev/**PROJECT_ID-redacted**/image-repo/
myimage@sha256:e4637b83784b96cb...93349a8864cc9b8be1 \
--show-provenance
```

A truncated example output is provided here. Note that the numbers are added for illustration purposes – they are not in the command output:

```
image_summary:
...
provenance_summary:
  provenance:
  - build:
...
[1]     materials:
        - uri: https://github.com/**REPO-redacted**/packt-
cloudbuild/commit/497985408c6f656944dd8d2f2ffd42160c59334e
...
```

This is the location of the source materials – in this case, the repository and specific commit:

```
[2]     subject:
        - digest:
          sha256:
e4637b83784b96cbdcaad40cde0622bd3c0e0bfe4cd4c993349a8864cc9b8
be1
          name: https://us-central1-docker.pkg.dev/**PROJECT_
ID-redacted**/image-repo/myimage
...
```

See the following container image SHA and location:

```
[3] envelope:
        payload: eyJfdHlwZSI6Imh0dHBzOi...
mZkNDIxNjBjNTkzMzRlIn1dfX0=
        payloadType: application/vnd.in-toto+json
        signatures:
        - keyid: projects/verified-builder/locations/global/
keyRings/attestor/cryptoKeys/builtByGCB/cryptoKeyVersions/1
          sig: MEUCIQDkyICeKz0Assa6Dgj4ss-rsd0RJ2ZXJWmBI-6lycoV_
wIgZnSaPYCRAGRo0BVYGx_3aofMJS0YifWLcNtwXQpmtQ0=
...
```

The critical pieces, which are the encoded envelope, signature, and public key, to verify that the contents of the metadata have not been tampered with or manipulated, are as follows:

- `payload` – The JSON format of the build metadata and build steps adhering to the SLSA spec
- `keyid` – The location of the `builtByGCB` public key to be used to validate the signature
- `sig` – The signature generated by Cloud Build for validation of the payload

This may repetition of information that is available in the Cloud Build logs, but it provides organizations with another mechanism to audit and ensures that a container image has not been tampered with or manipulated.

Attestations

GCP uses an attestation in the form of a digital document as a way to certify a container image. Attestations can be added to build steps to ensure that the steps were completed by a designated system. Cloud Build can generate an attestation that a build was completed in Cloud Build and this can be verified by the container runtime.

Binary Authorization is the service in GCP that container runtimes such as **Google Kubernetes Engine** (**GKE**) and Cloud Run use to determine whether an image can be deployed. For instance, GKE can deny an image for use because it was not built by Cloud Build. Organizations can generate an attestation that the image has passed QA. Using Binary Authorization, an organization can only permit container images that have passed QA to be deployed into the runtime. Additional attestations can be added as build steps defined for Cloud Build.

When Binary Authorization is enabled, an attestor is created by Cloud Build, which is then used to create the `built-by-cloud-build` attestation and signed using the `builtByGCB` crypto keys. This is validated by the supported container runtimes before the specified container image is deployed. *Figure 6.5* shows how Cloud Build can automatically sign images with a `built-by-cloud-build` attestation, and in the QA pipeline, how a configured build step can also sign a `validated-by-qa` custom attestation:

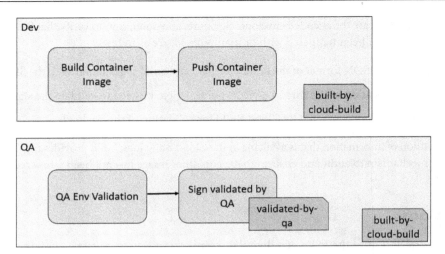

Figure 6.5 – Cloud Build signing attestations

In the following example, the `enforcementMode` for Binary Authorization is set to DRYRUN_
AUDIT_LOG_ONLY, which means the runtime will check for attestations, but only log violations. If
GKE (as the container runtime) enables Binary Authorization, when a Pod is requested to be created,
it won't be rejected but the following will be logged:

```
imagepolicywebhook.image-policy.k8s.io/dry-run: "true"
imagepolicywebhook.image-policy.k8s.io/overridden-
verification-result: "'nginx' : Image nginx denied by attestor
projects/**PROJECT_ID-redacted**/attestors/built-by-cloud-
build: Expected digest with sha256 scheme, but got tag or
malformed digest
```

In the preceding example, a generic `nginx` container image was requested to be deployed but was
caught and logged by Binary Authorization. Binary Authorization can be set up to require multiple
attestations as it passes through different parts of the build process, such as QA.

Figure 6.6 shows Binary Authorization checking for attestations in GKE's admission controller. If
all the required attestations are present and `enforcementMode` is ENFORCED_BLOCK_AND_
AUDIT_LOG, it is permitted into the cluster. This blocks admission of the container image if the
required attestations are not available:

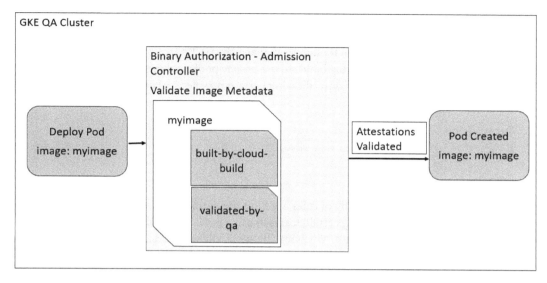

Figure 6.6 – A GKE container runtime using Binary Authorization
to determine whether an image can be admitted

The combination of both Cloud Build and GCP container runtimes supporting Binary Authorization can help improve the security posture of an organization. Denying container images that have not passed through the security rigor within an organization is an added layer of defense that can be implemented. Binary Authorization can also be configured to allow the deployment of images without the appropriate attestations in the event of an emergency. This mechanism is referred to as breakglass, which will bypass the validation.

Cloud Build's data has multipurpose uses for auditing and validation throughout your end-to-end pipeline. Specific GCP services can take advantage of the metadata, such as signatures, to determine whether resources have completed all required steps in the respective environment or not.

Securing the network perimeter

Cloud Build can leverage a GCP security construct named VPC Service Controls (`https://cloud.google.com/vpc-service-controls`) to guard against data exfiltration. **VPC Service Controls (VPC SC)** allows an organization to set policies to define the user information, service accounts, IP addresses, and IP subnetworks required to access a GCP service.

In the context of Cloud Build, only builds using private pools, as discussed in *Chapter 2, Configuring Cloud Build Workers*, can support VPC SC. Private pool instances can leverage this capability because they are associated with your VPC, even though it's managed by GCP. Organizations can also restrict Cloud Build even further by only allowing builds to use private pools within an organization's policy. Further restrictions can be applied by only allowing certain private worker pools at various GCP hierarchies:

- Organization
- Folder
- Project

By leveraging this fine-grained hierarchy of rules, we limit the permitted worker pools at each level. This isolates things further by restricting certain teams to using pools based on the required resources. It would also deny the use of default pools, which may be applicable when organizations require builds to run within their private networks.

When a VPC SC perimeter is configured, only those granted access can trigger a build or call the Cloud Build API, for example. In the previous examples as part of our *Defense in depth* section, we focused on service account role bindings that prevent access to particular services and data. VPC SC provides another layer of protection at the network layer by configuring associated policies. To create an initial VPC SC perimeter, we specify Cloud Build as a restricted service within the perimeter. This would mean that the Cloud Build service is protected and any ingress or egress from the perimeter would have to be defined by an ingress or egress policy.

> **Note**
> Organizations just getting started may want to configure VPC SC in dry run mode to audit the impact of defining a perimeter. Enabling enforced mode immediately may cause builds to fail. Any perimeters created in dry run mode can be converted to enforced mode without having to re-write policies.

As an ingress policy, for example, we may want to allow the user-specified service account defined in the trigger to access the Cloud Build API. For inner-loop scenarios where we may want to give specific developers access to trigger a build, we can do this by creating an access-level policy that clears specific developers within a specific IP/CIDR range or country, or who are using devices that adhere to specific policies.

We have limited access at both the **Identity and Access Management** (**IAM**) and VPC network layer, by using VPC SC and the defined service perimeter. The service perimeter can be extended to allow communication from other connecting networks for build triggering and service access from on-premises systems or specific developer machines.

VPC SC provides an additional layer of security at the network layer for services on GCP. In this chapter, we discussed how Cloud Build can leverage this GCP service to protect how builds can be triggered, but so that resources that build configurations can also access them.

Summary

Cloud Build integrates with various GCP services to provide users with different security capabilities. These capabilities help provide layers to an organization's defense-in-depth strategy. Cloud Build makes it easy to implement security within your pipeline but the shared responsibility model between the GCP services and the organization is critical to ensuring an end-to-end security posture.

In the next chapter, we will start looking into leveraging Cloud Build for the automation of infrastructure resources.

Part 3:
Practical Applications

In this part of the book, we will focus on reviewing Cloud Build functionality in the context of practical applications, including secure CI/CD for GKE, managing safe rollouts of cloud infrastructure via infrastructure as code, building a workflow from a local source to production in Cloud Run, and capabilities that can be leveraged when running in a production environment. You will walk away with concrete workflows you can step through and implement in your own Cloud Build and GCP environments.

This part comprises the following chapters:

- *Chapter 7, Automating Deployment with Terraform and Cloud Build*
- *Chapter 8, Securing Software Delivery for GKE with Cloud Build*
- *Chapter 9, Automating Serverless with Cloud Build*
- *Chapter 10, Running Operations for Cloud Build in Production*

Automating Deployment with Terraform and Cloud Build

The flexibility of **Cloud Build** provides a foundation for automating more than just the pipeline for building applications and services. The capabilities inherent in Cloud Build, such as leveraging Git as the source of truth and container images to drive the execution of a build step, help formulate the automation of infrastructure provisioning. One such pattern for this is the combination of using Cloud Build and **Terraform** (`https://www.terraform.io/`). In this chapter, we will walk through an example of using Cloud Build to automatically provision application infrastructure using Terraform's configuration language known as **HashiCorp Configuration Language** (HCL) and the **Command-Line Interface** (CLI).

We will cover the following topics in this chapter:

- Treating infrastructure as code
- Building a custom builder
- Managing the principle of least privilege for builds
- Human-in-the-loop with manual approvals

Treating infrastructure as code

What does it mean to treat **infrastructure as code** (IaC)? Infrastructure can be provisioned in many ways, such as imperatively or manually executing a set of commands to bring up an environment. An IaC provider intends to define how you would like an environment or component to be set up using a series of configurations. As noted in the introduction, with Terraform, we will be using its HCL to define a standard configuration interpreted by the `terraform` CLI tool. This configuration dictates the type of provider, such as a plugin, to interact with the service's API and provision what is specified. The configuration files help define the resource you wish to create and are ideally stored in a **Source Code Management** (SCM) repository for version control management.

Building code requires a set of instructions and defined prerequisites necessary to successfully compile and package the code. This configuration is defined in the Cloud Build configuration, along with the supporting container images for each step and the configuration dependencies stored in an SCM repository. When working with infrastructure, particularly a cloud provider such as **Google Cloud**, we will also define a Cloud Build configuration, the supporting container image to execute the Terraform CLI, the Terraform configuration files, and the configuration dependencies stored in an SCM repository.

Cloud Build can help orchestrate the execution and processing of Terraform configurations using its CLI tool. Terraform configurations are intended to be idempotent, meaning that re-running the Terraform command using the same configuration after the first run should result in zero changes, assuming that the infrastructure configuration was not changed outside of the pipeline. This allows for Cloud Build trigger referencing and the SCM repository to execute multiple times and only change the infrastructure if the configurations have changed.

Each Cloud Build execution (as in, build) is stateless. If the state does need to be persisted, it can be done in build steps to a service such as **Google Cloud Storage** (GCS). Otherwise, overall, the supported artifacts in a build can be stored in a service such as **Artifact Registry**. In the case of Terraform, it also requires the state to be stored externally so it is aware of configurations that have been provisioned. If the state is stored only within the run, Terraform wouldn't be aware of previous actuations. For Google Cloud, it is recommended to store this state in a **GCS** bucket with version control enabled. This allows for the state to be maintained, versioned, protected with permissions, encrypted at rest, and optionally encrypted using encryption keys. An additional example tutorial can be found here: `https://cloud.google.com/architecture/managing-infrastructure-as-code`.

The examples in this chapter focus on using Terraform to provision Google Cloud compute and network services. It is worth noting that Terraform resources are also available for provisioning Cloud Build components, such as triggers and private worker pools.

Simple and straightforward Terraform

Let's get started by preparing a bucket in GCS to store our Terraform state.

> **Note**
> The examples in this chapter were executed through a GCP Cloud Shell environment – you may run these in your terminal, but the configuration of gcloud and other tools may be necessary.

Set the project that you would like to create this bucket in:

```
$ gcloud config set project ${PROJECT_ID}
$ PROJECT_ID=$(gcloud config get-value project)
```

Create the bucket and enable versioning:

```
$ gsutil mb gs://${PROJECT_ID}
$ gsutil versioning set on gs://${PROJECT_ID}
```

> **Note**
>
> By enabling the versioning capability, we are asking GCS to maintain older versions of objects while newer objects are saved. The bucket name must also be unique throughout Google Cloud, not just within your project.

The Terraform state (managed within GCS) is also different from the Terraform configuration (for example, .tf or .tfvars), which are typically stored in an SCM for version control.

Once we have our bucket set up, before we start using Cloud Build to automate our infrastructure pipeline, let's do a quick test to spin up a **Google Compute Engine (GCE) virtual machine (VM)**.

We'll start with some quick files to get Terraform started.

> **Note**
>
> The examples include syntax and versions that worked during the writing of this book. However, changes may need to be made depending on the version of the Terraform CLI tool and Terraform provider for Google Cloud.
>
> In this example, we are using the environment variable, PROJECT_ID, that we stored in the previous command to generate the following backend.tf file. Terraform does not allow us to use Terraform variables to dynamically specify the backend bucket used for storing the state.

We will reference the bucket that was created in the previous command, to tell terraform where to store and retrieve the state:

```
$ mkdir ~/packt/cloudbuild-terraform; cd ~/packt/cloudbuild-
terraform
cat << EOF > backend.tf
terraform {
  backend "gcs" {
    bucket        = "${PROJECT_ID}"
    prefix        = "tfstate"
  }
}
EOF
```

Specify the versions of terraform that will work with this configuration, as well as the Google Cloud provider version:

```
$ cat << EOF > versions.tf
terraform {
  required_version = ">= 1.0.0, < 1.2.0"
  required_providers {
    google = {
      source  = "hashicorp/google"
      version = ">= 3.68.0"
    }
  }
}
provider "google" {
  project     = var.project_id
  region      = var.region
  zone        = var.zone
}
EOF
```

The list of variables and their default values that are required for setting up the Google Cloud provider and the GCE VM instance we will be asking Terraform to spin up. The following example is broken up into different sections:

```
$ cat << EOF > variables.tf
variable "project_id" {
  description = "Unique identifer of the Google Cloud Project
that is to be used"
  type        = string
  default     = "${PROJECT_ID}"
}
variable "region" {
  description = "Google Cloud Region in which regional GCP
resources are provisioned"
  type        = string
  default     = "us-west1"
}
variable "zone" {
```

```
  description = "Google Cloud Zone in which regional GCP
resources are provisioned"
  type      = string
  default   = "us-west1-a"
}
EOF
```

Network-related variables, such as name and network range, will be used by the compute instance:

```
$ cat << EOF >> variables.tf
variable "network_name" {
  description = "The name of the network to be used when
deploying resources"
  type      = string
  default   = "packt-network"
}
variable "subnet_name" {
  description = "The name of the subnet to be used when
deploying resources"
  type      = string
  default   = "packt-subnet"
}
variable "subnet_cidr" {
  description = "The name of the subnet to be used when
deploying resources"
  type      = string
  default   = "10.128.0.0/24"
}
EOF
```

Here are the compute-related variables that define the machine type, name, and **operating system (OS)** image:

```
$ cat << EOF >> variables.tf
variable "instance_name" {
  description = "The name of the GCE VM instance named to be
provisioned"
  type      = string
  default   = "packt-instance"
```

```
}
variable "instance_machine_type" {
  description = "The machine type you want the GCE VM to use"
  type        = string
  default     = "e2-medium"
}
variable "instance_image" {
  description = "The OS image you want the GCE VM boot disk to
use"
  type        = string
  default     = "rocky-linux-cloud/rocky-linux-8-v20220406"
}
EOF
```

Set up the network to be used by the GCE VM instance:

```
$ cat << EOF > vpc.tf
resource "google_compute_network" "packt_network" {
  name                    = "vpc-network"
  auto_create_subnetworks = false
}
resource "google_compute_subnetwork" "packt_subnet" {
  name          = var.subnet_name
  ip_cidr_range = var.subnet_cidr
  region        = var.region
  network       = google_compute_network.packt_network.name
}
EOF
```

The bare-minimum configuration required to spin up a GCE VM instance is as follows:

```
$ cat << EOF > gce.tf
resource "google_compute_instance" "packt_instance" {
  name         = var.instance_name
  machine_type = var.instance_machine_type
  boot_disk {
    initialize_params {
      image = var.instance_image
```

```
    }
  }
  network_interface {
    subnetwork = google_compute_subnetwork.packt_subnet.name
  }
}
EOF
```

Now that we have all the Terraform configuration files defined, we can run a few Terraform commands to test and spin up the GCE instance we have configured. We start with the `terraform init` command to verify our connectivity to the backend and the necessary modules specified within our configurations:

```
$ terraform init
```

The following is an example output of the `init` command:

```
Initializing the backend...
Successfully configured the backend "gcs"! Terraform will
automatically
use this backend unless the backend configuration changes.
Initializing provider plugins...
- Reusing previous version of hashicorp/google from the
dependency lock file
- Installing hashicorp/google v4.16.0...
- Installed hashicorp/google v4.16.0 (signed by HashiCorp)
Terraform has been successfully initialized!
```

We run the `terraform plan` command to preview the changes specified in the configuration, along with the `-out` switch to output the plan as recommended by Terraform. This will allow for validation and code reviews of what Terraform will attempt to perform on the specified Google Cloud project. Developers and operators can use `terraform show terraform.tfplan` to view the plan:

```
$ terraform plan -out terraform.tfplan
```

The following is an example output of a plan:

```
Terraform used the selected providers to generate the following
execution plan. Resource actions are indicated
with the following symbols:
  + create
```

```
Terraform will perform the following actions:
  # google_compute_instance.main will be created
  + resource "google_compute_instance" "main" {
      + can_ip_forward      = false
      + cpu_platform        = (known after apply)
...

          + node_affinities {
              + key      = (known after apply)
              + operator = (known after apply)
              + values   = (known after apply)
            }
        }
    }
Plan: 1 to add, 0 to change, 0 to destroy.
...
Saved the plan to: terraform.tfplan
To perform exactly these actions, run the following command to
apply:
    terraform apply "terraform.tfplan"
```

Rather than just using `terraform apply` to actuate the changes, we now have a `plan` file to specifically apply the approved changes if this were in a pipeline:

$ terraform apply terraform.tfplan

The following is an example output of an `apply` function:

```
google_compute_instance.main: Creating...
google_compute_instance.main: Still creating... [10s elapsed]
google_compute_instance.main: Creation complete after 11s
[id=projects/**REDACTED_project_id**/zones/us-west1-a/
instances/test-instance]
Apply complete! Resources: 1 added, 0 changed, 0 destroyed.
```

Now that we've manually tested a Terraform execution, we can add it to a Cloud Build configuration pipeline to automate this process. In Cloud Build, we will execute the individual Terraform command steps described previously, with each represented as a build step. If you do not de-provision (as in, destroy) the previously created resources, the Cloud Build configuration will not make any changes, because the Terraform configurations created previously are being re-used. Run the following command to destroy the created resources:

```
$ terraform destroy --auto-approve
```

Before we start using Cloud Build, we will need to grant the Cloud Build default service account network and compute permissions to create our resources. The previous commands, when running `terraform` in the terminal, used your local service account. When we execute with Cloud Build and we do not specify a service account, it will use the Cloud Build default service account:

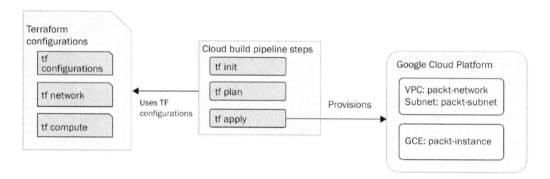

Figure 7.1 – Cloud Build and Terraform, all in one approach

First, we need to retrieve the Google Cloud project number, as it is needed to construct the Cloud Build default service account, PROJECT_NUMBER@cloudbuild.gserviceaccount.com:

```
$ export PROJECT_NUMBER=$(gcloud projects list \
    --filter="$(gcloud config get-value project)" \
    --format="value(PROJECT_NUMBER)"
)
```

Give the Cloud Build service account administrative permissions on the resources we will be creating:

```
$ gcloud projects add-iam-policy-binding ${PROJECT_ID} \
    --member=serviceAccount:${PROJECT_NUMBER}@cloudbuild.
gserviceaccount.com \
    --role=roles/compute.networkAdmin
```

```
$ gcloud projects add-iam-policy-binding ${PROJECT_ID} \
  --member=serviceAccount:${PROJECT_NUMBER}@cloudbuild.
gserviceaccount.com \
  --role=roles/compute.instanceAdmin.v1
```

In this example, we will be using the `hashicorp/terraform:1.0.0` container image to provide the `terraform` executable, starting with `terraform init`:

```
$ cat << EOF > cloudbuild.yaml
timeout: 3600s
steps:
- id: 'tf init'
  name: 'hashicorp/terraform:1.0.0'
  env:
  - "TF_IN_AUTOMATION=true"
  entrypoint: 'sh'
  args:
  - '-c'
  - |
      terraform init -input=false
EOF
```

The next build step is `terraform plan`, for outlining the changes:

```
$ cat << EOF >> cloudbuild.yaml
- id: 'tf plan'
  name: 'hashicorp/terraform:1.0.0'
  env:
  - "TF_IN_AUTOMATION=true"
  entrypoint: 'sh'
  args:
  - '-c'
  - |
      terraform plan -input=false -out terraform.tfplan
EOF
```

The last build step is `terraform apply`, for applying the outlined changes:

```
$ cat << EOF >> cloudbuild.yaml
- id: 'tf apply'
  name: 'hashicorp/terraform:1.0.0'
  env:
  - "TF_IN_AUTOMATION=true"
  entrypoint: 'sh'
  args:
  - '-c'
  - |
    terraform apply -input=false terraform.tfplan
EOF
```

Once we have our Cloud Build configuration available to us, we can manually execute the build by running the command used in previous chapters to submit a build. The default Cloud Build service account will be used:

```
$ gcloud builds submit --region us-west1
```

A successful build will yield the following truncated output with a SUCCESS status:

```
...
Step #2 - "tf apply": Apply complete! Resources: 3 added, 0
changed, 0 destroyed.
Finished Step #2 - "tf apply"
PUSH
DONE
...
STATUS: SUCCESS
```

> **Note**
>
> In automation, the `TF_IN_AUTOMATION` variable is set so that some instructions are truncated and manual instructions are not presented. For example, after executing `terraform plan` automatically with the variable set, the following statement is not outputted to the console: "`To perform exactly these actions...`".

If you recall from earlier, it was noted that Terraform is intended to be idempotent, so subsequent build submissions with the same configuration will yield the following output for a terraform plan execution:

```
Step #1 - "tf plan": No changes. Your infrastructure matches
the configuration.
Step #1 - "tf plan":
Step #1 - "tf plan": Your configuration already matches the
changes detected above. If you'd like
Step #1 - "tf plan": to update the Terraform state to match,
create and apply a refresh-only plan.
```

Executing `tf apply` to a plan with no changes will result in the following:

```
Step #2 - "tf apply": Apply complete! Resources: 0 added, 0
changed, 0 destroyed.
```

There are many things that we can do to build on top of this example, such as putting the created files into an SCM repository for version control and auditing. You can find this example at `https://github.com/PacktPublishing/Cloud-Native-Automation-With-Google-Cloud-Build/tree/main/chapter07/terraform` – the files that we created previously are located in the `single` folder.

The separation of resource creation and the build steps

In the following section, we will look into the `multiple` folder, where we expand on the example in the previous section by separating the creation of the VPC and subnet from the GCE VM instance. This will allow resource management with more separation and control. The configuration and setup for this are available in the `multiple` folder of the following repository: `https://github.com/PacktPublishing/Cloud-Native-Automation-With-Google-Cloud-Build/tree/main/chapter07/terraform`.

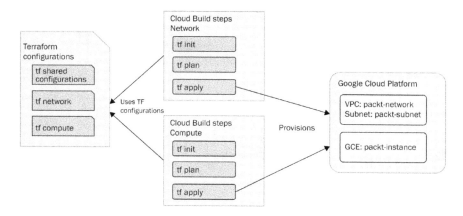

Figure 7.2 – Cloud Build and Terraform (VPC and compute provisioning as separate pipelines)

The majority of examples are the same, except for moving specific resource configurations into their respective vpc and compute folders. In the previous example, it consisted of executing the trio of terraform init, plan, and apply commands once – however, when the resources are separated, the trio of terraform commands will be executed twice, once for each set of resource configurations. Another difference is because they are executed separately, we will need to use another capability available to us in Terraform known as remote state.

remote state allows us to reference provisioned resources from other applied Terraform configurations, so rather than manually leveraging variables to reference a resource, we can retrieve the outputs of a previous Terraform execution to link to created resources. When we create a GCE VM instance, we need access to the subnet that the resource will be associated with.

Note the following output of the gce.tf configuration:

```
resource "google_compute_instance" "packt_instance" {
  name          = var.instance_name
  machine_type = var.instance_machine_type
  boot_disk {
    initialize_params {
      image = var.instance_image
    }
  }
  network_interface {
```

```
    subnetwork = data.terraform_remote_state.vpc.outputs.
subnet_name
  }
}
```

The subnetwork references `remote state` from our VPC configuration. The VPC Terraform configuration outputs the `subnet_name` value that will be used by our GCE VM.

Similar to the Terraform backend configuration, the `remote state` references the VPC state in the GCS bucket and path specified by the VPC backend.

```
data "terraform_remote_state" "vpc" {
  backend = "gcs"
  config = {
    bucket  = "${PROJECT_ID}"
    prefix  = "tfstate/vpc"
  }
}
```

In the pipeline, folders are used to separate the resources, so the Cloud Build steps need to know where to find the appropriate Terraform configurations – we use the `dir` notation to signify where to execute the `terraform` commands.

```
- id: 'tf init - vpc'
  name: 'hashicorp/terraform:1.0.0'
  dir: chapter07/terraform/multiple/vpc
  env:
  - "TF_IN_AUTOMATION=true"
  entrypoint: 'sh'
```

Let's go ahead and execute this Cloud Build infrastructure pipeline. While the configurations have changed and additional build steps were added, the result is the same as the single instance we constructed earlier in the chapter. If you would like to execute this yourself, after cloning the repository, you will need to replace some variables:

```
$ gcloud config set project ${PROJECT_ID}
$ PROJECT_ID=$(gcloud config get-value project)
$ envsubst < chapter07/terraform/multiple/variables.tf_tmpl >
multiple/variables.tf
$ envsubst < chapter07/terraform/multiple/vpc/backend.tf_tmpl >
multiple/vpc/backend.tf
```

```
$ envsubst < chapter07/terraform/multiple/compute/backend.tf_
tmpl > multiple/compute/backend.tf
$ envsubst < chapter07/terraform/multiple/compute/vpc-remote-
state.tf_tmpl > multiple/compute/vpc-remote-state.tf
```

Now, we can submit the Cloud Build configuration:

```
$ gcloud builds submit --region us-west1 --config cloudbuild-
multiple.yaml
```

Notice that the additional steps for the creation of VPC resources and compute resources are separate, unlike the simple example, Apply complete! Resources: 3 added, previously deployed:

```
Step #2 - "tf apply - vpc": Apply complete! Resources: 2 added,
0 changed, 0 destroyed.
Step #2 - "tf apply - vpc":
Step #2 - "tf apply - vpc": Outputs:
Step #2 - "tf apply - vpc":
Step #2 - "tf apply - vpc": subnet_name = "packt-m-subnet"
...
Step #5 - "tf apply - compute": Apply complete! Resources: 1
added, 0 changed, 0 destroyed.
Finished Step #5 - "tf apply - compute"
```

Just as with the simple preceding example, executing the multiple example over and over again will output the following for each of the resources attempting to be provisioned:

```
Apply complete! Resources: 0 added, 0 changed, 0 destroyed.
```

> **Note**
>
> The respective cloudbuild-destroy.yaml and cloudbuild-multiple-destroy.
> yaml are available in the repository to clean up the created resources. You will need to make adjustments to the variables, backends, and remote states to reference your project:
>
> ```
> $ gcloud builds submit --region us-west1 --config cloudbuild-
> multiple-destroy.yaml
> ```

The following example output shows the destroyed resources that were created in the multiple example created previously:

```
Step #2 - "tf apply - compute": Apply complete! Resources: 0
added, 0 changed, 1 destroyed.
```

```
. . .
Step #5 - "tf apply - vpc": Apply complete! Resources: 0 added,
0 changed, 2 destroyed.
```

The primary focus of the book has been the compilation of the source code stored in an SCM repository. The flexibility of Cloud Build allows each step to use a container image – in this case, we used the terraform image to deploy Terraform configurations using multiple build steps. So far, we've used pre-built images in our build steps and described the concept of custom builders. We will be jumping into the creation of our own custom builder in the next section.

Building a custom builder

Cloud Build uses a container image for each build step, which provides flexibility to essentially execute whatever binary or commands your build requires. Another pattern is where you build a custom builder image consisting of all the necessary build binaries required by the team. This can reduce complexity by having only one image to be used in build steps, but also only one image to be maintained.

Let's start with an example where your build needs access to a few Google Cloud resources such as gcloud and gsutil, but you would still like a few more for your overall builds such as terraform, kustomize, and skaffold. While the base image provides the first two, we would have to either use the available community builder images for each tool (https://github.com/GoogleCloudPlatform/cloud-builders-community), build a new custom image for each tool, or a single image that the team can maintain with the latter three.

We start with a Dockerfile that we define, the smaller cloud-sdk image based on alpine, and then add a few tools we desire, along with some cleanup:

```
$ mkdir ~/packt/cloudbuild-custombuilder; cd ~/packt/
cloudbuild- custombuilder
$ gcloud config set project ${PROJECT_ID}
$ PROJECT_ID=$(gcloud config get-value project)

$ cat << EOF > Dockerfile
FROM gcr.io/google.com/cloudsdktool/cloud-sdk:alpine
ENV TERRAFORM_VERSION=1.1.0
RUN echo "INSTALL TERRAFORM v${TERRAFORM_VERSION}" \
&& wget -q -O terraform.zip https://releases.hashicorp.com/
terraform/${TERRAFORM_VERSION}/terraform_${TERRAFORM_VERSION}_
linux_amd64.zip \
&& unzip terraform.zip \
&& chmod +x terraform \
```

```
&& mv terraform /usr/local/bin \
&& rm -rf terraform.zip
RUN gcloud components install \
kustomize \
skaffold \
&& rm -rf \$(find google-cloud-sdk/ -regex ".*/__pycache__") \
&& rm -rf google-cloud-sdk/.install/.backup
EOF
```

The preceding is just an example, but while we've primarily focused on Cloud Build as a mechanism to execute Cloud Build configurations, it can also build container images noted in previous chapters. Depending on your environment, you may need additional permissions, to enable Artifact Registry, and create a repository named core-image:

```
$ gcloud builds submit --region us-west1 --tag us-docker.pkg.
dev/${PROJECT_ID}/core-image/builder:latest
```

We can use the build image as described in the following Cloud Build configuration file:

```
$ cat << EOF > cloudbuild.yaml
steps:
- id: 'See what gcloud components are installed'
  name: 'us-docker.pkg.dev/${PROJECT_ID}/core-image/
builder:latest'
  entrypoint: 'sh'
  args:
  - '-c'
  - |
    gcloud components list --only-local-state --format
"value(id)"
- id: 'Check terraform version'
  name: 'us-docker.pkg.dev/${PROJECT_ID}k/core-image/
builder:latest'
  entrypoint: 'sh'
  args:
  - '-c'
  - |
    terraform version
EOF
```

Executing a manual `builds submit` of the preceding configuration will install `gcloud` components and validate the `terraform` version available:

```
$ gcloud builds submit --region us-west1
```

The following is a sample output of which `gcloud` components are installed locally for the image and the version of `terraform`:

```
Step #0 - "See what gcloud components are installed": Your
current Google Cloud CLI version is: 381.0.0
Step #0 - "See what gcloud components are installed": kubectl
Step #0 - "See what gcloud components are installed": bq
Step #0 - "See what gcloud components are installed": core
Step #0 - "See what gcloud components are installed": kustomize
. . .
Step #1 - "Check terraform version": Terraform v1.1.0
Step #1 - "Check terraform version": on linux_amd64
. . .
```

The team that owns the image may differ depending on your organization, but ideally, it is maintained by a single team. In an end-to-end build, requirements, suggestions, and feedback may come from different teams within the organization. Versioning becomes critical as well when certain builds depend on specific versions and combinations of tools that need to be considered when maintaining a custom builder image. It comes down to what works for the organization but the flexibility of the requirements of different teams can lead to optimal outcomes.

Managing the principle of least privilege for builds

In the previous chapter, we covered the underlying principle of least privilege for our build pipelines. While the example in the first section of this chapter leveraged the Cloud Build default service account, it was convenient, but depending on the type of pipeline or automation we want to run, we may not want to provide a service account that has the ability to manipulate both the network and compute. One way to achieve this is by separating our Cloud Build pipeline configurations – we can also minimize the impact of mistakes or the attack surface.

If you haven't cloned the repo, go ahead and clone it (https://github.com/PacktPublishing/Cloud-Native-Automation-With-Google-Cloud-Build):

```
$ git clone https://github.com/PacktPublishing/Cloud-Native-Automation-With-Google-Cloud-Build
```

Navigate to this chapter's example:

```
$ cd Cloud-Native-Automation-With-Google-Cloud-Build/chapter07/
terraform
```

In the Cloud Build configuration, `cloudbuild-multiple-network.yaml`, we specify the specific `serviceAccount` that is to be used by this build:

```
$ serviceAccount: 'projects/${PROJECT_ID}/
serviceAccounts/${CLOUDBUILD_NETWORK}@${PROJECT_ID}.iam.
gserviceaccount.com'
```

We will first need to prepare the variables to be used by both our Cloud Build and Terraform configurations:

```
$ gcloud config set project ${PROJECT_ID}
$ export PROJECT_ID=$(gcloud config get-value project)
$ export CLOUDBUILD_NETWORK=cloudbuild-network
$ export CLOUDBUILD_COMPUTE=cloudbuild-compute
```

Create our two service accounts and assign the permissions to be used by the respective Terraform configurations. Create the account that will build out our VPC:

```
$ gcloud iam service-accounts create ${CLOUDBUILD_NETWORK} \
   --display-name="Cloud Build Network Admin"
$ gcloud projects add-iam-policy-binding ${PROJECT_ID} \
   --member=serviceAccount:${CLOUDBUILD_NETWORK}@${PROJECT_ID}.
iam.gserviceaccount.com \
   --role=roles/compute.networkAdmin
$ gcloud projects add-iam-policy-binding ${PROJECT_ID} \
   --member=serviceAccount:${CLOUDBUILD_NETWORK}@${PROJECT_ID}.
iam.gserviceaccount.com \
   --role=roles/iam.serviceAccountUser
$ gcloud projects add-iam-policy-binding ${PROJECT_ID} \
   --member=serviceAccount:${CLOUDBUILD_NETWORK}@${PROJECT_ID}.
iam.gserviceaccount.com \
   --role=roles/logging.logWriter
$ gcloud projects add-iam-policy-binding ${PROJECT_ID} \
   --member=serviceAccount:${CLOUDBUILD_NETWORK}@${PROJECT_ID}.
iam.gserviceaccount.com \
   --role=roles/storage.objectAdmin
```

Next, we create the account that will provision our GCE VM instance:

```
$ gcloud iam service-accounts create ${CLOUDBUILD_COMPUTE} \
  --display-name="Cloud Build Compute Admin"
$ gcloud projects add-iam-policy-binding ${PROJECT_ID} \
  --member=serviceAccount:${CLOUDBUILD_COMPUTE}@${PROJECT_ID}.
iam.gserviceaccount.com \
  --role=roles/compute.instanceAdmin.v1
$ gcloud projects add-iam-policy-binding ${PROJECT_ID} \
  --member=serviceAccount:${CLOUDBUILD_COMPUTE}@${PROJECT_ID}.
iam.gserviceaccount.com \
  --role=roles/iam.serviceAccountUser
$ gcloud projects add-iam-policy-binding ${PROJECT_ID} \
  --member=serviceAccount:${CLOUDBUILD_COMPUTE}@${PROJECT_ID}.
iam.gserviceaccount.com \
  --role=roles/logging.logWriter
$ gcloud projects add-iam-policy-binding ${PROJECT_ID} \
  --member=serviceAccount:${CLOUDBUILD_COMPUTE}@${PROJECT_ID}.
iam.gserviceaccount.com \
  --role=roles/storage.objectAdmin
```

Permissions may take a few moments to propagate – submit the build and see the following:

```
ERROR: (gcloud.builds.submit) INVALID_ARGUMENT: could
not resolve source: googleapi: Error 403: cloudbuild-
compute@**REDACTED_project_id**.iam.gserviceaccount.com does
not have storage.objects.get access to the Google Cloud Storage
object., forbidden
```

We will need to substitute a few variables in our Cloud Build configurations. In this example, we specify the service account in the configuration because we will manually be submitting this over the CLI. If you use a trigger, you will be able to specify the service account as part of the trigger as well:

```
$ envsubst < cloudbuild-multiple-network.yaml_tmpl >
cloudbuild-multiple-network.yaml
$ envsubst < cloudbuild-multiple-compute.yaml_tmpl >
cloudbuild-multiple-compute.yaml
$ envsubst < cloudbuild-multiple-network-destroy.yaml_tmpl >
cloudbuild-multiple-network-destroy.yaml
$ envsubst < cloudbuild-multiple-compute-destroy.yaml_tmpl >
cloudbuild-multiple-compute-destroy.yaml
```

Once the variables have been replaced, we should be ready to go with submitting our build configurations. Just as in the previous scenario where we submitted the network and compute builds separately, we will be doing it separately here too. The main difference in this scenario is we are using a service account that has the necessary privileges to perform the task at hand:

```
$ gcloud builds submit --region us-west1 --config cloudbuild-
multiple-network.yaml
```

> **Note**
> Because we are using a user-specified service account and we specified it using
> CLOUD_LOGGING_ONLY, we won't be able to see the logs stream in the terminal. At the time
> of writing, with gcloud beta, you could stream the logs in your terminal. Alternatively, the
> output from the gcloud command is an option to click a **Logs are available at** link to open
> the console to view it over the UI. If you have another terminal window open, you can view
> the logs using gcloud builds log YOUR_BUILD_ID.

Given that each service account has specific resource permissions, if you were to replace serviceAccount in the cloudbuild-multiple-compute.yaml configuration with the cloudbuild-network service account, it would not have the appropriate permissions to create a compute instance. We would end up with the following error:

```
Step #2 - "tf apply - compute": google_compute_instance.packt_
instance: Creating...
Step #2 - "tf apply - compute": |
Step #2 - "tf apply - compute": | Error: Error creating
instance: googleapi: Error 403: Required 'compute.instances.
create' permission for 'projects/**REDACTED-project_id**/zones/
us-west1-a/instances/packt-m-instance'
Step #2 - "tf apply - compute": | More details:
Step #2 - "tf apply - compute": | Reason: forbidden,
Message: Required 'compute.instances.create' permission for
'projects/**REDACTED-project_id**/zones/us-west1-a/instances/
packt-m-instance'
Step #2 - "tf apply - compute": | Reason: forbidden,
Message: Required 'compute.disks.create' permission for
'projects/**REDACTED-project_id**/zones/us-west1-a/disks/packt-
m-instance'
```

Leveraging the principle of least privilege helps determine which accounts are able to manipulate their respective resources, allowing for the life cycle of each account to be managed independently. As noted earlier, it can also reduce the attack surface by limiting permissions to specific resources.

Human-in-the-loop with manual approvals

Just as in code, when dealing with infrastructure-related builds, there may be a need to have a human intervene and approve it before a build can begin. By separating our resources into different build configurations as in the previous section, we can determine which resources would require approval before the build pipeline can execute.

> **Note**
>
> For this current section, we will walk through an example – it is not intended for you to copy and paste, as your setup and configurations may differ.

In this example, we will be using the publicly hosted GitHub repository that is connected to a GCP project. We need to do this because the approval mechanism is only available through triggers. Assuming you have already connected to your repository via the console (i.e., GitHub (Cloud Build GitHub App)) or a supported SCM provider of choice, you can use the following CLI to create a trigger.

This example is available at `https://github.com/PacktPublishing/Cloud-Native-Automation-With-Google-Cloud-Build/tree/main/chapter07/terraform` using the `cloudbuild-multiple-network-approval.yaml` Cloud Build configuration file. The following are the example commands for this section. The repository is provided as a starting point, but your fork or repository layout may be different:

```
$ gcloud beta builds triggers create github \
    --name=packt-cloudbuild-terraform-approval \
    --repo-name= Cloud-Native-Automation-With-Google-Cloud-Build \
    --repo-owner=PacktPublishing \
    --branch-pattern=^main$ \
    --require-approval \
    --build-config=/chapter07/terraform/cloudbuild-multiple-network-approval.yaml \
    --service-account "projects/**REDACTED-project_id**/serviceAccounts/cloudbuild-network@**REDACTED-project_id**.iam.gserviceaccount.com"
```

The console is also available for creating triggers as well, as noted in a previous chapter.

Once the trigger has been created and a build has been triggered through an SCM commit or manually **RUN** in Cloud Build, the build will be in a wait status until it is approved or rejected. Note the **APPROVE** and **REJECT** options in this screenshot of the GCP Console UI:

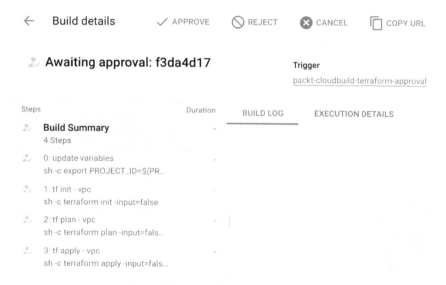

Figure 7.3 – A build awaiting approval

If the **APPROVE** option is selected, a dialog will prompt the user for approval confirmation.

> **Note**
> For a user to approve or reject the build, they must have the **Cloud Build Approver** role assigned to their account.

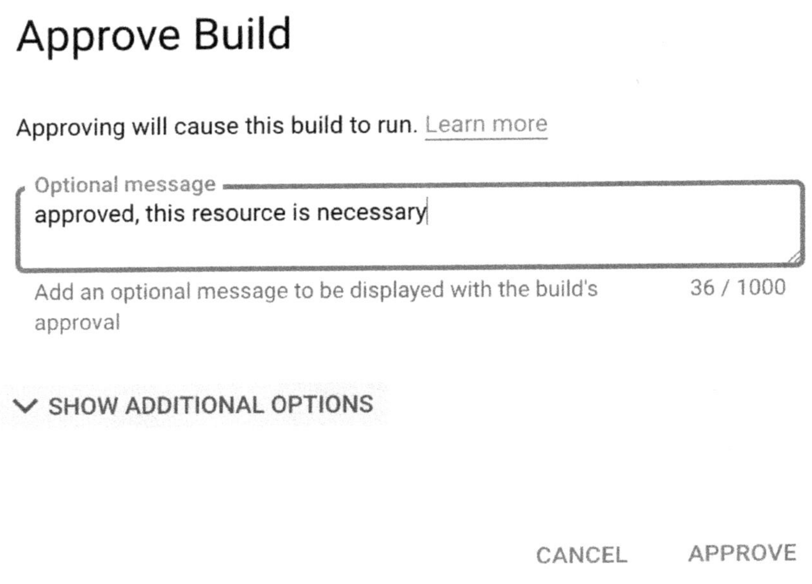

Approve Build

Approving will cause this build to run. Learn more

Optional message ──
approved, this resource is necessary|

Add an optional message to be displayed with the build's   36 / 1000
approval

∨ SHOW ADDITIONAL OPTIONS

CANCEL   APPROVE

Figure 7.4 – The approval dialog

Once the **APPROVE** option has been selected in the preceding dialog, the build will begin. This can also be performed using the gcloud CLI. Using the CLI, the user will need to know the build ID to approve or reject:

```
$ gcloud alpha builds approve f3da4d17-d832-449a-b925-
9ae95245f5b0 \
  --project=**REDACTED-project_id**
```

This is a sample output when approval is performed using the CLI:

```
metadata:
  '@type': type.googleapis.com/google.devtools.cloudbuild.
v1.BuildOperationMetadata
  ...
    result:
      approvalTime: '2022-04-14T23:45:41.598449Z'
      ...
      decision: APPROVED
    state: APPROVED
```

Manual approval allows a human or service to intervene to determine whether or not a build should be executed. As time passes, it may become more viable to have a system validate and make a decision or eventually remove the approval requirement altogether. The intent is to provide a level of flexibility to the organization.

Summary

Cloud Build is not just for building and compiling services and applications. The flexibility of the platform, consisting of pre-built cloud builders, custom cloud builders, user-specified service accounts, and human approvals, can help automate tasks within an organization. In this chapter, leveraging Terraform, we were able to provision resources on GCP using IaC concepts.

In the next chapter, we will dive into the deep end of securing the delivery of your services within **Google Kubernetes Engine (GKE)**.

8

Securing Software Delivery to GKE with Cloud Build

Software supply chain security has become a critical focus of the industry in recent years, with numerous compromises resulting in damaging outcomes for companies and users alike. This focus has yielded more rigorous practices in securing software delivery, underscoring the importance of practices such as verifying trust in the artifacts you deploy and applying the **principle of least privilege** (**POLP**). With Cloud Build being a recommended mechanism for automating software delivery in Google Cloud, it is important to understand its capabilities and best practices around security.

In this chapter, we will walk through an example of leveraging Cloud Build to deploy a set of applications to a private **Google Kubernetes Engine** (**GKE**) cluster while implementing multiple security best practices using Cloud Build features.

Specifically, we will cover these topics in this chapter:

- Creating your build infrastructure and deployment target
- Securing build and deployment infrastructure
- Applying POLP to builds
- Configuring release management for builds
- Enabling verifiable trust in artifacts from builds

Creating your build infrastructure and deployment target

The deployment target in this example is a GKE cluster. Cloud Build will implement our software delivery pipeline as a build, performing both the building of container images and the release of Kubernetes manifests to run the container images in the GKE cluster.

Before we dive into how you can secure this software delivery to GKE from Cloud Build, we should review the aspects of Kubernetes and GKE that are relevant to our software delivery processes.

Kubernetes (`https://kubernetes.io`) is a popular **open source software** (**OSS**) project owned by the **Cloud Native Computing Foundation** (**CNCF**) (`https://www.cncf.io`) that is responsible for running containerized applications across numerous machines.

At a high level, it achieves this via software that acts as a control plane coordinating the lifecycle of containerized applications across a set of servers registered with the control plane called **nodes**. These nodes can be shared by multiple containerized applications or can be isolated for specific applications. In addition, Kubernetes also allows for the logical separation of applications via a Kubernetes resource called a **namespace**.

Together, these components make up a Kubernetes cluster, as depicted in the following diagram:

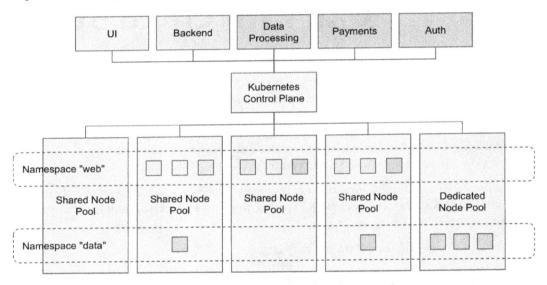

Figure 8.1 – High-level architecture of a Kubernetes cluster

The control plane is critical for this chapter's examples as it is the foundational point of interaction with the cluster for users and automation alike. It provides a network endpoint exposing the Kubernetes **application programming interface** (**API**) and runs software that performs functions such as the admission of containerized applications into the cluster.

This means that the interactions with the control plane must be secure and that we must utilize Cloud Build security functionality with its capabilities for admission control.

If you are new to Kubernetes and would like to dive deeper, it is recommended to visit the official documentation (`https://kubernetes.io/docs`) to learn more about how it operates.

GKE is a platform in which Google provides managed Kubernetes clusters as a service to users. The level of management that GKE provides depends on the cluster mode you utilize. You can choose from the following cluster modes:

- **GKE Standard**, in which the control plane is fully managed by Google but you can still control aspects of node provisioning and upgrade lifecycle

- **GKE Autopilot**, in which both the control plane and node lifecycles are fully managed by Google

In this example, you will be using GKE to create a GKE Standard cluster, which means that the control plane will run in a Google-owned environment and be fully managed by Google:

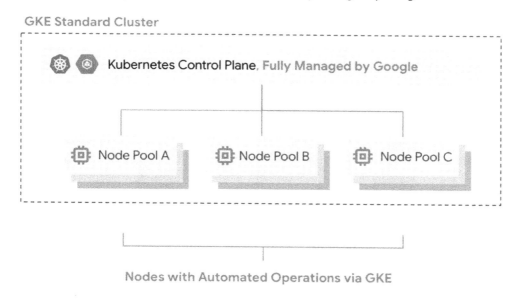

Figure 8.2 – High-level architecture of a GKE Standard cluster

Specifically, you will create a *private* GKE Standard cluster. This is a configuration for a GKE cluster in which the control plane is only accessible by a private endpoint on a network managed by Google and peered with your own Google **virtual private cloud** (**VPC**).

This means that we can align Cloud Build and GKE security best practices by using private pools in Cloud Build to build and deploy to private GKE clusters. This ensures that Cloud Build's calls to the Kubernetes control plane do not travel across the public internet, but rather over a private network connection.

Enabling foundational Google Cloud services

To get started with this example and create a secure build infrastructure, you must first ensure that you enable all relevant services in your Google Cloud project.

It is highly recommended that you run this chapter's examples in a sandbox or test Google Cloud project, as you will have full control not only to run the example but also to easily clean up resources created.

You will begin by initiating a session in Cloud Shell (`shell.cloud.google.com`), the ephemeral Linux workstation provided by Google Cloud.

Clone the example repository from GitHub using the following command:

```
$ cd ~ && git clone https://github.com/agmsb/cloudbuild-gke.git
&& cd cloudbuild-gke/08
```

Run the `gcloud` command to configure `gcloud` to use the appropriate Google Cloud project as follows, replacing `<add-project-id>` with the Google Cloud project you want to use:

```
$ gcloud config set project <add-project-id>
```

Given that it is recommended that this example be run in a sandbox Google Cloud project, it is best completed with the role of a project owner assigned to your user in **Identity and Access Management (IAM)**.

With the project owner role, you can enable the services required for this project, using the following command:

```
$ gcloud services enable container.googleapis.com
cloudbuild.googleapis.com artifactregistry.googleapis.
com binaryauthorization.googleapis.com servicenetworking.
googleapis.com containeranalysis.googleapis.com
```

This command will enable the following services in your project:

- GKE
- Cloud Build
- Artifact Registry
- Binary Authorization
- Service Networking API
- Container Analysis API

With these services enabled, you can now begin setting up the resources for this chapter's example.

Setting up the VPC networking for your environment

We will run numerous commands that will require names for various resources; for your convenience, default names are provided in a file accessible via the GitHub repository.

It is recommended you walk through this example using these variable names; once you have completed it, you can go back and reuse the example and change the variable names to something with more meaning specific to your use case.

Source the `bin/variables.sh` file to read the environment variables into your current session, like so:

```
$ source bin/variables.sh
```

With the proper variables set, we can move forward with creating our first set of resources, as follows:

- A VPC for your GKE cluster
- A VPC with which your fully managed private pool workers can peer
- Cloud **virtual private network** (**VPN**) tunnels connecting the aforementioned VPCs over a private, encrypted connection

Change directories into the `vpc` directory, in which you will find a `main.tf` file. This contains Terraform configuration code to create the preceding resources. Here's the code you need to execute:

```
$ cd infra/vpc
```

To use the Terraform configuration code, you will need to initialize Terraform in that directory by running the following command using the `terraform` **command-line interface** (**CLI**). This was last tested on version `1.2.7` of `terraform`:

```
$ terraform init
```

Review all of the resources to be created by the `terraform` CLI using the following command:

```
$ terraform plan
```

To actualize those changes, run `terraform apply`, as follows. You will need to also type in "`yes`" when prompted in the terminal; this allows you to proceed with the provisioning of the resources in `main.tf`:

```
$ terraform apply
```

This will take a number of minutes to complete. Once the VPC networking resources for our example have been created, we can move forward with creating a GKE cluster.

Setting up your private GKE cluster

Because GKE is a managed Kubernetes offering, it provides numerous turnkey features enabled by parameters passed to GKE upon cluster creation.

As a reminder, if you have lost context since working on the previous section, run the following commands to navigate to the right directory and source the proper environment variables.

Replace `<add-project-id>`, shown here, with the project you are working with:

```
$ gcloud config set project <add-project-id>
```

Navigate to the directory you are working in, like so:

```
$ cd ~/cloudbuild-gke/08
```

Source the required environment variables by running the following command:

```
$ source bin/variables.sh
```

Now, we can create a GKE cluster. Run the following command to kick off cluster creation. Do note that this consumes quota from **Google Compute Engine** (**GCE**) and Persistent Disk, creating a cluster with six nodes in total, each with 20 **gigabytes** (**GB**) of Persistent Disk:

```
$ gcloud container clusters create $CLUSTER_NAME \
    --project=$PROJECT_ID \
    --cluster-version=$CLUSTER_VERSION \
    --enable-ip-alias \
    --network=$CLUSTER_VPC_NAME \
    --subnetwork=$CLUSTER_SUBNET_NAME \
    --cluster-secondary-range-name=pod \
    --services-secondary-range-name=svc \
    --region=$REGION \
    --num-nodes=2 \
    --disk-size=20 \
    --enable-private-nodes \
    --enable-private-endpoint \
    --master-ipv4-cidr=$CLUSTER_CONTROL_PLANE_CIDR \
    --workload-pool=$PROJECT_ID.svc.id.goog \
    --enable-binauthz
```

There are a few things worth noting in this cluster configuration. The first is that we are able to lock down the cluster to only private networking via the `-enable-private-nodes`, `-enable-private-endpoint`, and `-master-ipv4-cidr` parameters.

The `-enable-private-nodes` parameter ensures that no nodes running in the GKE cluster have a public **Internet Protocol** (**IP**) address. This by default blocks the nodes from pulling any publicly hosted container images at locations such as Docker Hub. By configuring your cluster in such a manner, you are implicitly denying these images from running in your cluster—a good start for ensuring your cluster runs known and trusted software.

Later in this chapter, in the *Enabling verifiable trust in artifacts from builds* section, we will review how to explicitly ensure you run known and trusted software using the last parameter we specified: `-enable-binauthz`.

Given this configuration, this also means we cannot pull from Container Registry or Artifact Registry via public IP addresses. To still ensure that our cluster can run containers hosted in Google Cloud registries, we will need Private Services Access. This is a feature in Google Cloud that enables both Google and third parties to offer services accessible via a private IP address rather than public endpoints.

By default, this is enabled for GKE private clusters; however, if you are looking to replicate this example with a shared VPC, you will need to manually enable Private Services Access. You can find more information on how to do this here: `https://cloud.google.com/vpc/docs/private-google-access`.

The `-enable-private-endpoint` parameter ensures that we only make our GKE control plane available via a private IP address. This means that API calls from humans and automation alike do not traverse the public internet. These API calls may have sensitive information that leaks details about your software, such as container image names, environment variables, or secrets. Thus, it is important to ensure that these calls are made privately.

Finally, `--master-ipv4-cidr=$CLUSTER_CONTROL_PLANE_CIDR` defines the range that Google will use to allocate private IP addresses for you to interact with the control plane. This is because the control plane is actually owned and run by Google; the way it interacts with your nodes is via a peering that is created with your VPC, which is where your GKE nodes and other resources you create reside.

While this peering is created for you automatically upon cluster creation, we actually need to update this peering to do the following:

- Export custom routes to connect to the private pool.

- Decline the exchanging of routes for public IP destinations.

Run the following commands to extract the name of the automatically created peering, and update it according to the preceding points:

```
$ export GKE_PEERING_NAME=$(gcloud container clusters describe
$CLUSTER_NAME \
    --region=$REGION \
    --format='value(privateClusterConfig.peeringName)')
$ gcloud compute networks peerings update $GKE_PEERING_NAME \
```

```
--network=$CLUSTER_VPC_NAME \
--export-custom-routes \
--no-export-subnet-routes-with-public-ip
```

Let's review the current architecture of the GKE cluster, now that we have created the VPC for the GKE private cluster and the cluster itself. You can view this here:

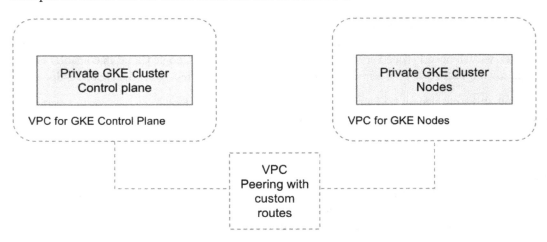

Figure 8.3 – High-level architecture of the private GKE cluster created

To summarize, the critical aspects of what you have created include the following:

- Ensuring access to the GKE control plane is only available via a private connection.
- Nodes cannot pull container images from public registries.
- Nodes pull container images via private, internal IP addresses.

With this completed, let's now move forward with creating a secure build and deployment infrastructure using Cloud Build private pools.

Securing build and deployment infrastructure

When securing your software delivery processes in Cloud Build, it is important to begin with the underlying infrastructure that runs the builds themselves. If compromised, attackers can access sensitive information such as your source code, your secrets that the build may access, and the deployment targets with which your builds interact.

Underlying infrastructure in this chapter's example specifically means the following:

- Private pool workers that execute Cloud Build builds

- VPC networking, connecting workers to systems such as Artifact Registry and GKE

- Minimal or managed container images executing build steps

We will begin by creating a private pool for our example.

Creating private pools with security best practices

Previously introduced in *Chapter 2, Configuring Cloud Build Workers*, private pools are a specific mode for Cloud Build workers that have distinct features from the default pool, including VPC connectivity and greater machine-type availability, all while remaining fully managed in a Google-owned environment.

We will be using a private pool in this example to run four builds, as follows:

- Team A, Build 1—Build a container image for a Python app.

- Team A, Build 2—Deploy the container image to GKE.

- Team B, Build 1—Build a container image for a Python app.

- Team B, Build 2—Deploy the container image to GKE.

This simulates a scenario in which two different teams both deploy their workloads to the same Kubernetes cluster configured for simple multi-tenancy by isolating the applications in the cluster using Kubernetes namespaces, as illustrated in the following diagram:

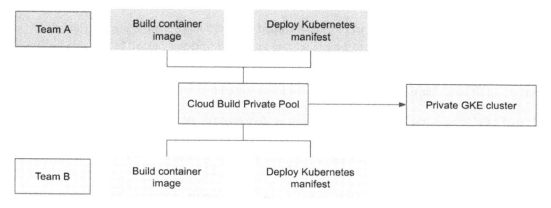

Figure 8.4 – High-level architecture of the software delivery workflow in this chapter

Before we can run the builds, we must create our private pool and set it up to communicate with our GKE cluster, all with the following goals in mind:

- Create secure workers to run builds in the private pool.

- Securely access the GKE control plane.

- Limit access to the GKE control plane via allowlists.

- Secure the network perimeter in which Cloud Build and GKE communicate.

As a reminder, if you have lost context since working on the previous section, run the following commands to navigate to the right directory and source the proper environment variables.

Replace <add-project-id>, shown here, with the project you are working with:

```
$ gcloud config set project <add-project-id>
```

Navigate to the directory you are working in, as follows:

```
$ cd ~/cloudbuild-gke/08
```

Source the required environment variables, like so:

```
$ source bin/variables.sh
```

Beginning with the first goal in mind, you will create a Cloud Build private pool and ensure that the security configurations available to us are utilized. But before we can create a private pool, there are a couple of extra networking resources we have to create, as follows:

- Private IP range for private pool workers

- VPC peering to peer the VPC we previously created with the Service Networking API (https://cloud.google.com/service-infrastructure/docs/enabling-private-services-access), enabling connectivity to private pool workers

To create an IP range for the private pool, run the following command:

```
$ gcloud compute addresses create $PRIVATE_POOL_IP_RANGE_NAME \
      --global \
      --addresses=$PRIVATE_POOL_IP_RANGE \
      --purpose=VPC_PEERING \
      --prefix-length=$PRIVATE_POOL_IP_RANGE_SIZE \
      --network=$PRIVATE_POOL_VPC_NAME
```

To create a peering with the VPC we created earlier, run the following commands:

```
$ gcloud services vpc-peerings connect \
    --service=servicenetworking.googleapis.com \
    --ranges=$PRIVATE_POOL_IP_RANGE_NAME \
    --network=$PRIVATE_POOL_VPC_NAME \
    --project=$PROJECT_ID
$ gcloud compute networks peerings update servicenetworking-
googleapis-com \
    --network=$PRIVATE_POOL_VPC_NAME \
    --export-custom-routes \
    --no-export-subnet-routes-with-public-ip
```

We created the peering with two configurations of note: `--export-custom-routes` and `--export-custom-routes`. These configurations respectively allow for us to export routes between our private pool and our private GKE control plane, while also preventing any routes with public IP addresses to be exported.

With our peering properly set up, we will finish our networking configuration by advertising our private pool and GKE control-plane IP addresses, as follows:

```
$ gcloud compute routers update-bgp-peer ${PRIVATE_POOL_ROUTER}
\
    --peer-name=$PRIVATE_POOL_ROUTER_PEER_0 \
    --region=${REGION} \
    --advertisement-mode=CUSTOM \
    --set-advertisement-ranges=${PRIVATE_POOL_IP_
RANGE}/${PRIVATE_POOL_IP_RANGE_SIZE}

$ gcloud compute routers update-bgp-peer ${PRIVATE_POOL_ROUTER}
\
    --peer-name=$PRIVATE_POOL_ROUTER_PEER_1 \
    --region=${REGION} \
    --advertisement-mode=CUSTOM \
    --set-advertisement-ranges=${PRIVATE_POOL_IP_
RANGE}/${PRIVATE_POOL_IP_RANGE_SIZE}

$ gcloud compute routers update-bgp-peer ${CLUSTER_ROUTER} \
    --peer-name=${CLUSTER_ROUTER_PEER_0} \
    --region=${REGION} \
```

```
      --advertisement-mode=CUSTOM \
      --set-advertisement-ranges=$CLUSTER_CONTROL_PLANE_CIDR


$ gcloud compute routers update-bgp-peer ${CLUSTER_ROUTER} \
      --peer-name=${CLUSTER_ROUTER_PEER_1} \
      --region=${REGION} \
      --advertisement-mode=CUSTOM \
      --set-advertisement-ranges=$CLUSTER_CONTROL_PLANE_CIDR
```

Now, we can proceed with creating a private pool. To create your private pool, begin by sourcing the required environment variables and creating a private pool configuration file called `private-pool.yaml` under the `infra/` directory, as follows:

```
$ cat > infra/private-pool.yaml <<EOF
privatePoolV1Config:
  networkConfig:
    # egressOption: NO_PUBLIC_EGRESS
    peeredNetwork: projects/$PROJECT_ID/global/
networks/$PRIVATE_POOL_VPC_NAME
  workerConfig:
    machineType: e2-standard-2
    diskSizeGb: 100
EOF
```

The most important parts with regard to security in this configuration file are under the `networkConfig` stanza. Private pools, given that they are fully managed by Google, operate similarly to GKE control planes in that they are given access to resources in your VPC via VPC peering. You can see an illustration of this in the following diagram:

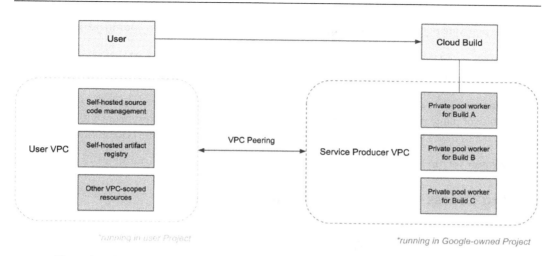

Figure 8.5 – Network architecture between your VPC and the VPC where private pools run

Let's review the egressOption and peeredNetwork fields and what they mean for our private pool, as follows:

- Setting egressOption to NO_PUBLIC_EGRESS removes public IP addresses from our workers in the private pool.

- Setting peeredNetwork to a distinct VPC separate from our GKE cluster VPC is required. This is due to a lack of transitive peering functionality.

If we were to remove public IP addresses from our private pool, this means that we could also deny our worker infrastructure access to the public internet and artifacts that could be either malicious or vulnerable. This helps reduce the surface area where your software delivery pipeline could be compromised.

> **Note**
>
> For this example, we will leave the NO_PUBLIC_EGRESS option out of our private pool config because the source repository to which we will be connecting will be GitHub.com. This is for simplicity and to avoid any need for licensing.
>
> In a secure software delivery setup, you would ideally utilize this parameter in conjunction with a private **source code management (SCM)** system such as GitHub Enterprise, which the private pool would be able to access over a private connection.

Now, create a private pool using the configuration file created in the previous command, like so:

```
$ gcloud builds worker-pools create $PRIVATE_POOL_NAME
--config-from-file infra/private-pool.yaml --region $REGION
```

However, for many, it is not enough to merely make communications between the build system and the GKE control plane private. Should an attacker have access to creating resources within a VPC that has access to that VPN, they may still be able to interact with your private GKE control plane.

Let's review how we can secure the control plane from other resources within our network perimeter.

Securing access to your private GKE control plane

As a reminder, if you have lost context since working on the previous section, run the following commands to navigate to the right directory and source the proper environment variables.

Replace <add-project-id>, shown here, with the project you are working with:

```
$ gcloud config set project <add-project-id>
```

Navigate to the directory you are working in, as follows:

```
$ cd ~/cloudbuild-gke/08
```

Source the required environment variables, like so:

```
$ source bin/variables.sh
```

It is recommended to configure an allowlist using the authorized networks for the GKE control-plane access feature in GKE (https://cloud.google.com/kubernetes-engine/docs/how-to/authorized-networks), as illustrated in the following screenshot:

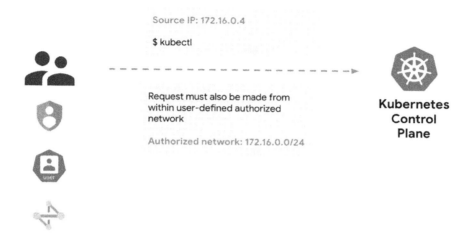

Figure 8.6 – Ensuring that users come from a known and trusted network identity

This will enable you to restrict access to the control plane by denying any traffic incoming from network identities that is not on a specific allowlist. This also means you must provide GKE with the network range that should be authorized to call the control plane. You can do this by running the following commands:

```
$ gcloud container clusters update $CLUSTER_NAME \
    --enable-master-authorized-networks \
    --region=$REGION \
    --master-authorized-networks=$PRIVATE_POOL_IP_
RANGE/$PRIVATE_POOL_IP_RANGE_SIZE
```

This now means that you will only be able to utilize Cloud Build workers from this private pool to interact with your GKE control plane. If you wanted to create a private workstation or a bastion host using a Compute Engine **virtual machine** (**VM**), you would also need to add its private IP address to the allowlist in order to interact with the GKE control plane using kubectl.

With the worker, cluster, and networking configuration set up, it is time to begin evaluating the permissions with which our builds run in this environment.

Applying POLP to builds

When builds run and interact with other Google Cloud resources, by default they utilize the Cloud Build service account as their identity, to which their permissions are assigned.

This, however, does not enable builds or users to apply POLP; if you have one GCP service account that is used by multiple builds that perform different tasks or interact with different resources, then all builds that use that service account will be overly privileged.

Cloud Build has support for per-build or per-trigger service accounts; this enables you to create each service account with intention and according to POLP.

This principle ensures that each build has no more permissions than the minimal amount it requires to execute successfully; this is achieved with purpose-specific GCP service accounts.

Creating build-specific IAM service accounts

Let's begin by creating two GCP service accounts, with one for each team we simulate in this example.

As a reminder, if you have lost context since working on the previous section, run the following commands to navigate to the right directory and source the proper environment variables.

Replace `<add-project-id>`, shown here, with the project you are working with:

```
$ gcloud config set project <add-project-id>
```

Navigate to the directory you are working in, as follows:

```
$ cd ~/cloudbuild-gke/08
```

Source the required environment variables, like so:

```
$ source bin/variables.sh
```

Now, create the first service account. This will be associated with Team A and will be associated with the build responsible for building a container image and pushing it to Team A's repository in Artifact Registry. Here's the code you need to execute:

```
$ gcloud iam service-accounts create $GCP_SA_NAME_01 \
    --display-name=$GCP_SA_NAME_01
```

The permission that this service account would require is the ability to write a container image to a specific repository in Artifact Registry. This adheres to POLP, as we do not want to grant access to *all* repositories in Artifact Registry, especially if we have other teams building and storing images that have nothing to do with Team A. You can see a representation of this in the following screenshot:

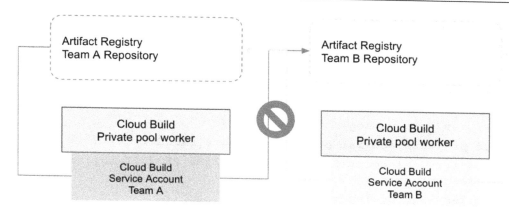

Figure 8.7 – Per-build service accounts with permissions to a specific Artifact Registry repository only

Let's create a repository in Artifact Registry along with a role binding that will give the service account permission to a single, named repository for Team A in Artifact Registry, as follows:

```
$ gcloud artifacts repositories create $REPOSITORY_A \
    --repository-format=docker \
    --location=$REGION
$ gcloud artifacts repositories add-iam-policy-binding team-a-
repository\
    --location us-west1 --member=$GCP_SA_NAME_01 --role= roles/
artifactregistry.writer
```

Let's now repeat the same process, except for Team B's specific service account for building and pushing container images to their own repository in Artifact Registry, like so:

```
$ gcloud iam service-accounts create $GCP_SA_NAME_02 \
    --display-name=$GCP_SA_NAME_02
$ gcloud artifacts repositories create $REPOSITORY_B \
    --repository-format=docker \
    --location=$REGION
$ gcloud artifacts repositories add-iam-policy-binding team-b-
repository\
    --location us-west1 --member=$GCP_SA_NAME_02 --role= roles/
artifactregistry.writer
```

We must also grant the service accounts logging access to write build logs to Cloud Logging, using the two following commands:

```
$ gcloud projects add-iam-policy-binding ${PROJECT_ID} \
--member="serviceAccount:${GCP_SA_NAME_01}@${PROJECT_ID}.iam.
gserviceaccount.com" \
--role=roles/logging.logWriter
$ gcloud projects add-iam-policy-binding ${PROJECT_ID} \
--member="serviceAccount:${GCP_SA_NAME_02}@${PROJECT_ID}.iam.
gserviceaccount.com" \
--role=roles/logging.logWriter
```

With this completed, we will now move on to configuring minimal permissions for the service accounts on the GKE cluster side. Similar to what we did with Artifact Registry, we will apply POLP to cluster access.

Custom IAM roles for build service accounts

Because Team A and Team B will have their own namespaces in your GKE cluster, we only want their service accounts to have permissions to those namespaces, rather than the entire GKE cluster, as depicted in the following diagram:

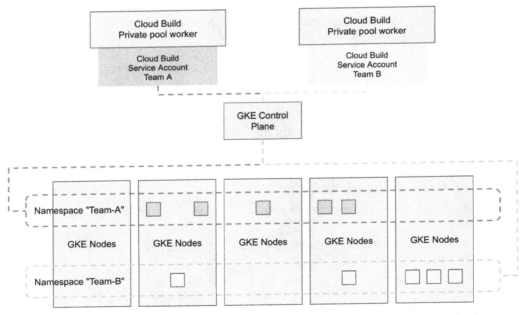

Figure 8.8 – GCP service accounts with namespace-level permissions via Workload Identity

To achieve this, we must create namespaces in your GKE cluster, custom IAM roles for the build service accounts for Team A and Team B, and utilize Kubernetes **role-based access control (RBAC)** to have minimal permissions within the GKE cluster.

For the purposes of disambiguation between very similarly named resources, see the following diagram:

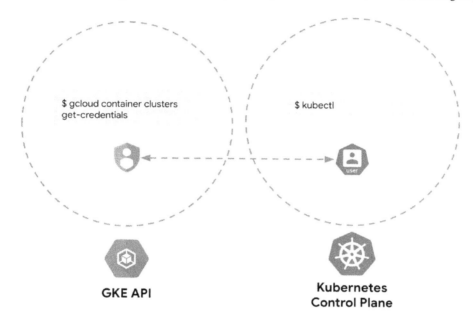

Figure 8.9 – GCP service account to fetch credentials; Kubernetes service account for everything else

Begin by creating a custom role that each service account will use to fetch credentials for the GKE cluster, as follows:

```
$ cat << EOF > infra/minimal-gke-role.yaml
title: minimal-gke
description: Gets credentials only, RBAC for authz.
stage: GA
includedPermissions:
- container.apiServices.get
- container.apiServices.list
- container.clusters.get
- container.clusters.getCredentials
$ gcloud iam roles create minimal-gke-role --project=$PROJECT_
ID\
    --file=infra/minimal-gke-role.yaml
```

Now, bind that custom role to each of the service accounts that will be used for each of the builds that deploy Team A and Team B's respective containerized applications to the GKE cluster, like so:

```
$ gcloud projects add-iam-policy-binding ${PROJECT_ID} \
--member="serviceAccount:${GCP_SA_NAME_01}@${PROJECT_ID}.iam.
gserviceaccount.com" \
--role=projects/$PROJECT_ID/roles/minimal_gke_role
$ gcloud projects add-iam-policy-binding ${PROJECT_ID} \
--member="serviceAccount:${GCP_SA_NAME_02}@${PROJECT_ID}.iam.
gserviceaccount.com" \
--role=projects/$PROJECT_ID/roles/minimal_gke_role
```

With your GCP service accounts set up with the custom role to get credentials from GKE clusters, we now must set up the Kubernetes resources. Run the following commands to create namespaces and Kubernetes service accounts in your cluster:

```
$ gcloud projects add-iam-policy-binding ${PROJECT_ID} \
--member="serviceAccount:${GCP_SA_GCB_DEFAULT}" \
--role=roles/container.admin
$ cat << EOF > infra/bootstrap-cluster.yaml
steps:
  - name: gcr.io/cloud-builders/gcloud
    id: Bootstrap GKE cluster
    entrypoint: bash
    args:
      - '-c'
      - |
        gcloud container clusters get-credentials cluster
--region $REGION --project $PROJECT_ID
        kubectl create ns $NAMESPACE_01
        kubectl create ns $NAMESPACE_02
        kubectl create role team-a-admin
--verb=get,list,watch,create,update,patch,delete
--resource=deployments.apps,services --namespace=$NAMESPACE_01
        kubectl create role team-b-admin
--verb=get,list,watch,create,update,patch,delete
--resource=deployments.apps,services --namespace=$NAMESPACE_02
        kubectl create rolebinding team-a-gcp-sa-binding
--role=team-a-admin --user=${GCP_SA_NAME_01}@${PROJECT_ID}.iam.
gserviceaccount.com --namespace=$NAMESPACE_01
```

```
        kubectl create rolebinding team-b-gcp-sa-binding
--role=team-b-admin --user=${GCP_SA_NAME_02}@${PROJECT_ID}.iam.
gserviceaccount.com --namespace=$NAMESPACE_02
options:
  workerPool: 'projects/$PROJECT_NUM/locations/$REGION/
workerPools/$PRIVATE_POOL_NAME'
EOF
$ gcloud builds submit . --config=infra/bootstrap-cluster.yaml
```

Let's now validate that both the GKE and Kubernetes resources have been set up accordingly. We will create a test build in which we will attempt to get Kubernetes deployments from each of the team namespaces we created.

Run the following commands to create and run the test build for Team A. The build should fail, but logs should reveal that this service account was able to get resources for the team-a namespace while failing to obtain resources for the team-b namespace:

```
$ cat << EOF > bin/test-build-a.yaml
steps:
  - name: gcr.io/cloud-builders/gcloud
    id: Test GCP SA for Team A
    entrypoint: bash
    args:
      - '-c'
      - |
        gcloud container clusters get-credentials $CLUSTER_NAME
--region $REGION --project $PROJECT_ID
        kubectl get deployments -n $NAMESPACE_01
        kubectl get deployments -n $NAMESPACE_02
serviceAccount: 'projects/$PROJECT_ID/serviceAccounts/${GCP_SA_
NAME_01}@${PROJECT_ID}.iam.gserviceaccount.com'
options:
  workerPool: 'projects/$PROJECT_NUM/locations/$REGION/
workerPools/private-pool'
  logging: CLOUD_LOGGING_ONLY
EOF


$ gcloud builds submit . --config=bin/test-build-a.yaml
```

The output in your Cloud Build logs should look similar to this:

```
>2022-07-01T15:02:23.226364190Z No resources found in team-a
namespace.

>2022-07-01T15:02:23.625289198Z Error from server (Forbidden):
deployments.apps is forbidden: User "build-01-sa@agmsb-lab.
iam.gserviceaccount.com" cannot list resource "deployments"
in API group "apps" in the namespace "team-b": requires one of
["container.deployments.list"] permission(s).
```

You can also test this out with the Team B service account if you so choose.

Having now secured the build infrastructure and ensured that the permissions with which builds run will be minimal, it is time to begin setting up the build execution.

Configuring release management for builds

In *Chapter 5, Triggering Builds*, we introduced build triggers; in this example, we will utilize triggers with `GitHub.com`, though in theory this workflow can be applied to other integrations such as GitHub Enterprise.

The workflow in this example will follow this ordering of steps:

1.   The developer works on a feature branch.
2.   The developer pushes code changes to the GitHub repository on that feature branch.
3.   The developer opens a **pull request** (**PR**) to the `main` branch, triggering the first build.
4.   Cloud Build runs tests and builds a container image for an app.
5.   The GitHub repository owner merges PRs, triggering the second build.
6.   Cloud Build runs and deploys the container image to the private GKE cluster.

The workflow will remain the same for both Team A and Team B. In order to get started with walking through this workflow, begin by setting up `GitHub.com` to integrate with Cloud Build.

Integrating SCM with Cloud Build

To run this example requires a `GitHub.com` account. You can begin by logging in to your GitHub account in the browser (`https://www.github.com`). If you do not have one, you can sign up for free here (`https://docs.github.com/en/get-started/signing-up-for-github/signing-up-for-a-new-github-account`).

You will also need to authenticate using the `gh` command-line utility built into Cloud Shell, which we will use to automate the creation of repositories for Team A and Team B.

As a reminder, if you have lost context since working on the previous section, run the following commands to navigate to the right directory and source the proper environment variables.

Replace `<add-project-id>`, as shown, with the project you are working with:

```
$ gcloud config set project <add-project-id>
```

Navigate to the directory you are working in, as follows:

```
$ cd ~/cloudbuild-gke/08
```

Source the required environment variables, like so:

```
$ source bin/variables.sh
```

Now, run the following command to step through the GitHub authentication flow and follow the prompts you receive:

```
$ gh auth login
```

Begin by creating and pushing two repositories to `GitHub.com`, one for Team A and one for Team B. We will use pre-built repository templates that contain the respective files for Team A and Team B.

Create and clone the Team A repository using the following commands:

```
$ cd bin
$ gh repo create $GH_A -public
$ gh repo clone $GH_A && cd $GH_A
```

Seed the empty repository with the repository template for Team A by running the following command:

```
$ cp -r ../../repo_templates/team_a/. .
```

Now, commit the changes and push them to the upstream repository, as follows:

```
$ git add .
$ git commit -m "Copy over example repo."
$ git push --set-upstream origin HEAD
```

We'll repeat the steps for Team B, creating and cloning the Team B repository using the following command:

```
$ cd ..
$ gh repo create $GH_B --public
$ gh repo clone $GH_B && cd $GH_B
```

Seed the empty repository with the repository template for Team B by running the following command:

```
$ cp -r ../../repo_templates/team_b/. .
```

Now, commit the changes and push them to the upstream repository, as follows:

```
$ git add .
$ git commit -m "Copy over example repo."
$ git push --set-upstream origin HEAD
```

Before we set up the triggers to kick off the builds from events from these repositories, we will need to install the GitHub App for Cloud Build to integrate with GitHub.com (https://cloud.google.com/build/docs/automating-builds/build-repos-from-github#installing_gcb_app).

> **Note**
>
> For this example, we are utilizing GitHub.com-based integrations with Cloud Build for simplicity and to avoid any requirement for licensing.
>
> In a more realistic example, you would want to set up integrations with a private SCM system such as GitHub Enterprise, with which Cloud Build would have private connectivity.
>
> To view how you can set up integrations such as this, view the documentation here: https://cloud.google.com/build/docs/automating-builds/build-repos-from-github-enterprise.

Now, create two triggers for Team A, like so:

```
$ gcloud beta builds triggers create github \
    --name=team-a-build \
    --region=$REGION \
    --repo-name=$GH_A \
    --repo-owner=$GH_USERNAME \
    --pull-request-pattern=master --build-config=build-a.yaml \
    --service-account=projects/$PROJECT_ID/
```

```
serviceAccounts/${GCP_SA_NAME_01}@${PROJECT_ID}.iam.
gserviceaccount.com
$ gcloud beta builds triggers create github \
    --name=team-a-deploy \
    --region=$REGION \
    --repo-name=$GH_A \
    --repo-owner=$GH_USERNAME \
    --branch-pattern=master --build-config=deploy-a.yaml \
    --service-account=projects/$PROJECT_ID/
serviceAccounts/${GCP_SA_NAME_01}@${PROJECT_ID}.iam.
gserviceaccount.com \
    --require-approval
```

Let's review some of the differences between the two triggers.

In the first trigger created, we are utilizing Cloud Build's support for triggering on PRs by using the `--pull-request-pattern` parameter. By specifying `main`, this means that builds will run upon PR being opened against the `master` branch.

In the second trigger created, we specify `-branch-pattern` instead to trigger on any changes to the `master` branch.

Now, create triggers for Team B, which will follow the same pattern as the triggers for Team A, as follows:

```
$ gcloud beta builds triggers create github \
    --name=team-b-build \
    --region=$REGION \
    --repo-name=$GH_B \
    --repo-owner=$GH_USERNAME \
    --pull-request-pattern=master --build-config=build-b.yaml \
    --service-account=projects/$PROJECT_ID/
serviceAccounts/${GCP_SA_NAME_02}@${PROJECT_ID}.iam.
gserviceaccount.com
$ gcloud beta builds triggers create github \
    --name=team-b-deploy \
    --region=$REGION \
    --repo-name=$GH_B\
    --repo-owner=$GH_USERNAME \
    --branch-pattern=master --build-config=deploy-b.yaml \
    --service-account=projects/$PROJECT_ID/
serviceAccounts/${GCP_SA_NAME_02}@${PROJECT_ID}.iam.
```

```
gserviceaccount.com \
    --require-approval
```

With the triggers now created, we can now review how we can gate these builds, ensuring that there can be additional human review before the builds actually run.

Gating builds with manual approvals

This brings us to the last difference you may have noticed between each of the two triggers for each team—the addition of the `--require-approval` flag, which introduces this method of explicit and manual approval for builds.

This provides a basis in which an administrator with the distinct IAM permissions of Cloud Build Approver can approve builds that require approval. Without this approval, the builds sit and wait and do not execute.

For this example workflow, we place manual approvals in front of the second trigger for both Team A and Team B, allowing for a gate prior to simulating a rollout to production.

If you visit the Cloud Build console, you will see the following options for a build awaiting approval:

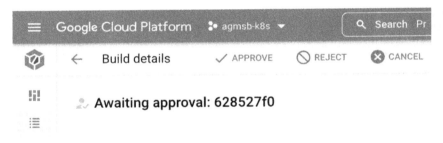

Figure 8.10 – Viewing individual build approval requests

When actually approving the build, not only does this action get captured in Cloud Audit Logging to understand who performed which rollout and when, but it also enables the Cloud Build Approver to provide reasons and metadata that defend the decision to approve the build, as illustrated in the following screenshot:

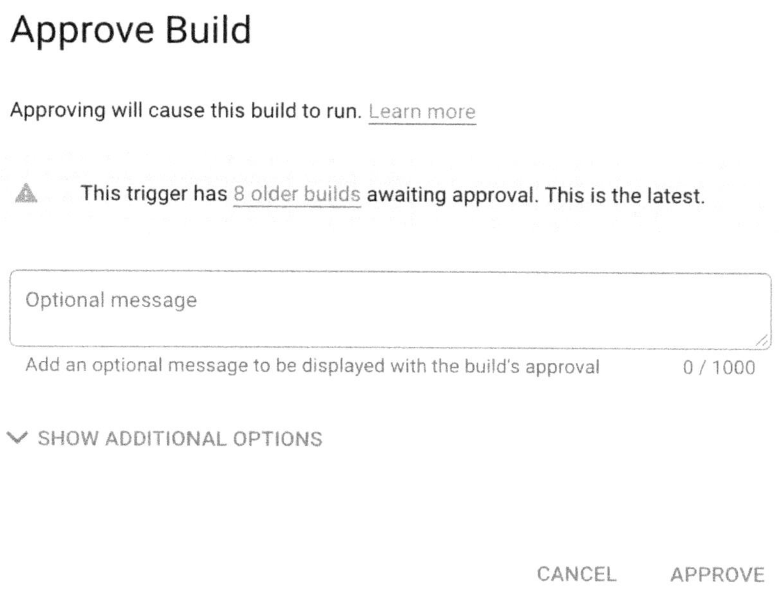

Approve Build

Approving will cause this build to run. Learn more

⚠ This trigger has 8 older builds awaiting approval. This is the latest.

Optional message

Add an optional message to be displayed with the build's approval   0 / 1000

∨ SHOW ADDITIONAL OPTIONS

CANCEL   APPROVE

Figure 8.11 – Viewing build approval dialog from the Cloud Build console

This allows for multiple gates, both implicit and explicit, that allow for human review to decide whether or not code and artifacts get promoted into different environments, including production.

Now, let's run through this example from end to end by triggering all four of the builds we described for Team A and Team B respectively.

Executing builds via build triggers

Begin by creating Cloud Build configuration files for each of these triggers.

As a reminder, if you have lost context since working on the previous section, run the following commands to navigate to the right directory and source the proper environment variables.

Replace <add-project-id>, shown here, with the project you are working with:

```
$ gcloud config set project <add-project-id>
```

Navigate to the directory you are working in, as follows:

```
$ cd ~/cloudbuild-gke/08
```

Source the required environment variables, like so:

```
$ source bin/variables.sh
```

Start by creating build files for building container images for Team A and Team B. These builds will be responsible for building a container image and pushing it to the Team A and Team B repositories in Artifact Registry respectively. Here's the code you need to execute:

```
$ cat << EOF > bin/$GH_A/build-a.yaml
steps:
  - name: gcr.io/cloud-builders/docker
    id: Build container image
    args: ['build', '-t', '${REGION}-docker.pkg.dev/$PROJECT_
ID/$REPOSITORY_A/team-a-app:v1',  '.']
images: [${REGION}-docker.pkg.dev/$PROJECT_ID/$REPOSITORY_A/
team-a-app:v1]
serviceAccount: 'projects/${PROJECT_ID}/serviceAccounts/${GCP_
SA_NAME_01}@${PROJECT_ID}.iam.gserviceaccount.com'
options:
  requestedVerifyOption: VERIFIED
  workerPool: 'projects/$PROJECT_NUM/locations/$REGION/
workerPools/$PRIVATE_POOL_NAME'
  logging: CLOUD_LOGGING_ONLY
images: [${REGION}-docker.pkg.dev/$PROJECT_ID/$REPOSITORY_A/
team-a-app:v1]
EOF


$ cat << EOF > bin/build-b.yaml
steps:
  - name: gcr.io/cloud-builders/docker
    id: Build container image
    args: ['build', '-t', '${REGION}-docker.pkg.dev/$PROJECT_
ID/$REPOSITORY_B/team-b-app:v1',  '.']
images: [${REGION}-docker.pkg.dev/$PROJECT_ID/$REPOSITORY_B/
team-b-app:v1]
serviceAccount: 'projects/${PROJECT_ID}/serviceAccounts/${GCP_
SA_NAME_02}@${PROJECT_ID}.iam.gserviceaccount.com'
options:
  requestedVerifyOption: VERIFIED
```

```
  workerPool: 'projects/$PROJECT_NUM/locations/$REGION/
workerPools/$PRIVATE_POOL_NAME'
  logging: CLOUD_LOGGING_ONLY
images: [${REGION}-docker.pkg.dev/$PROJECT_ID/$REPOSITORY_B/
team-b-app:v1]
EOF
```

Next, create build files to deploy Team A and Team B's applications to their respective namespaces in the GKE cluster, as follows:

```
$ cat << EOF > bin/$GH_A/deploy-a.yaml
steps:
  - name: gcr.io/cloud-builders/gke-deploy
    id: Prep k8s manifests
    args:
      - 'prepare'
      - '--filename=k8s.yaml'
      - '--image=${REGION}-docker.pkg.dev/$PROJECT_
ID/$REPOSITORY_A/team-a-app:v1'
      - '--version=v1'
  - name: gcr.io/google.com/cloudsdktool/cloud-sdk
    id: Get kubeconfig and apply manifests
    entrypoint: sh
    args:
      - '-c'
      - |
        gcloud container clusters get-credentials $CLUSTER_NAME
--region $REGION --project $PROJECT_ID
        kubectl apply -f output/expanded/aggregated-resources.
yaml -n $NAMESPACE_01
serviceAccount: 'projects/${PROJECT_ID}/serviceAccounts/${GCP_
SA_NAME_01}@${PROJECT_ID}.iam.gserviceaccount.com'
options:
  workerPool: 'projects/$PROJECT_NUM/locations/$REGION/
workerPools/$PRIVATE_POOL_NAME'
  logging: CLOUD_LOGGING_ONLY
EOF


$ cat << EOF > bin/$GH_B/deploy-b.yaml
```

```
steps:
  - name: gcr.io/cloud-builders/gke-deploy
    id: Prep k8s manifests
    args:
      - 'prepare'
      - '--filename=k8s.yaml'
      - '--image=${REGION}-docker.pkg.dev/$PROJECT_
ID/$REPOSITORY_A/team-b-app:v1'
      - '--version=v1'
  - name: gcr.io/google.com/cloudsdktool/cloud-sdk
    id: Get kubeconfig and apply manifests
    entrypoint: sh
    args:
      - '-c'
      - |
        gcloud container clusters get-credentials $CLUSTER_NAME
--region $REGION --project $PROJECT_ID
        kubectl apply -f output/expanded/aggregated-resources.
yaml -n $NAMESPACE_02
serviceAccount: 'projects/${PROJECT_ID}/serviceAccounts/${GCP_
SA_NAME_02}@${PROJECT_ID}.iam.gserviceaccount.com'
options:
  workerPool: 'projects/$PROJECT_NUM/locations/$REGION/
workerPools/$PRIVATE_POOL_NAME'
  logging: CLOUD_LOGGING_ONLY
EOF
```

With the build files in place for each team's repository, let's now get those changes upstream. We'll create a test branch and push those to our remote repository on GitHub.com, like so:

```
$ cd bin/$GH_A
$ git branch test
$ git checkout test
$ git add .
$ git commit -m "Add Cloud Build YAML files."
$ git push --set-upstream origin HEAD
$ cd ../$GH_B
$ git branch test
```

```
$ git checkout test
$ git add .
$ git commit -m "Add Cloud Build YAML files."
$ git push --set-upstream origin HEAD
$ cd ../..
```

We can now visit the web **user interface** (**UI**) for `GitHub.com` in which we will be prompted to create a PR in each repository respectively to merge the changes added to the test branch.

Upon opening the PRs for each repository, we will see our first pair of triggers kick off. These should each build the container images for Team A and Team B respectively. By specifying the `images` field, we also ensure that they will be pushed by Cloud Build with build provenance automatically generated, as previously discussed in *Chapter 6, Managing Environment Security*.

Now, we can merge each of the PRs for each repository, which will kick off our second pair of triggers. Each of these builds run two steps: first, preparing the Kubernetes manifests by inserting digests for each of the container images, and second, applying those manifests to the GKE cluster.

Because approvals are required, you must approve each of these builds before they can kick off. Once you approve the builds and they complete successfully, navigate to **Workloads** in the GKE menu in the sidebar. We can now see that `team-a-app` and `team-b-app` have successfully been deployed. You can also navigate to **Services and Ingress** in the GKE menu in the sidebar to access the load balancer endpoint that fronts requests to your respective apps.

Enabling verifiable trust in artifacts from builds

Finally, in securing the delivery of software from source to GKE via Cloud Build, you will want to ensure that you can verify that the software artifacts running in your cluster were indeed built in a trusted environment—in this case, your Cloud Build workers.

Cloud Build provides automatic build provenance (`https://cloud.google.com/build/docs/securing-builds/view-build-provenance`), which enables Cloud Build to generate signed metadata for each container image it builds, proving that the artifact originated from a build in Cloud Build and not **out-of-band** (**OOB**) via a bad actor.

Building images with build provenance

Finally, in securing the delivery of software from source to GKE via Cloud Build, you will want to ensure that you can verify that the software artifacts running in your cluster were indeed built in a trusted environment—in this case, your Cloud Build workers. You can see an illustration of this here:

Figure 8.12 – Build provenance provides crypto-signed metadata attesting artifact origin

We can view this metadata by running the following command:

```
$ gcloud artifacts docker images describe \
$REGION-docker.pkg.dev/$PROJECT_ID/$REPOSITORY_A/team-a-app:v1 \
    --show-provenance
```

Now, we will take a look at how we can use this metadata in our deployment processes.

Utilizing Binary Authorization for admission control

With this metadata generated from the Cloud Build perspective, we can now secure our target environments by enabling Binary Authorization and creating policies that will deny container images from being run that do not have the build provenance generated by Cloud Build.

This enables you to have the means by which to generate verifiable trust in Cloud Build, and verify that generated trust at runtime in GKE or Cloud Run. The process is illustrated in the following diagram:

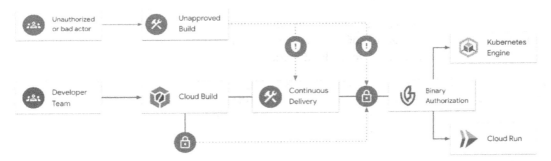

Figure 8.13 – Utilizing attestations to automatically admit or deny artifacts into the production GKE cluster

In this example, we will enable Binary Authorization by importing a policy that requires an attestation from the `built-by-cloud-build` attestor, which is automatically generated for you when running a build that generates build provenance.

Run the following command to create the required policy:

```
$ cat << EOF > infra/policy.yaml
globalPolicyEvaluationMode: ENABLE
defaultAdmissionRule:
    evaluationMode: REQUIRE_ATTESTATION
    enforcementMode: ENFORCED_BLOCK_AND_AUDIT_LOG
    requireAttestationsBy:
    - projects/${PROJECT_ID}/attestors/${ATTESTOR_ID}
EOF
```

Run the following command to import the required policy:

```
$ gcloud container binauthz policy import infra/policy.yaml
```

With this policy enabled, we now ensure that only images built by Cloud Build with build provenance will be admitted into our cluster at deployment time.

Let's test this out by trying to deploy a container image OOB, built locally in Cloud Shell and not via our workers in Cloud Build.

We'll do this by pulling an existing sample container not built by Cloud Build into our project. Configure Docker in Cloud Shell to pull images from Artifact Registry, and pull the `hello-app` image from `google-samples`, like so:

```
$ gcloud auth configure-docker $REGION-docker.pkg.dev
```

```
$ docker pull us-docker.pkg.dev/google-samples/containers/gke/
hello-app:1.0
```

Then, tag the container image as `out-of-band-app` and push it to Team A's repository in Artifact Registry, like so:

```
$ docker tag us-docker.pkg.dev/google-samples/containers/gke/
hello-app:1.0 $REGION-docker.pkg.dev/$PROJECT_ID/$REPOSITORY_A/
out-of-band-app:tag1
```

```
$ docker push $REGION-docker.pkg.dev/$PROJECT_ID/$REPOSITORY_A/
out-of-band-app:tag1
```

Now, let's create a Kubernetes manifest and apply it to our cluster using a build in Cloud Build, as follows:

```
$ cd bin
$ mkdir out-of-band && cd out-of-band

$ DIGEST=$(gcloud artifacts docker images describe $REGION-
docker.pkg.dev/$PROJECT_ID/$REPOSITORY_A/out-of-band-app:tag1
--format="json" | jq '.image_summary."digest"' | cut -d "\"" -f
2)

$ cat << EOF > k8s.yaml
apiVersion: apps/v1
kind: Deployment
metadata:
  name: out-of-band-app
spec:
  selector:
    matchLabels:
      app: out-of-band
  replicas: 1
  template:
    metadata:
      labels:
        app: out-of-band
    spec:
      containers:
      - name: app
        image: $REGION-docker.pkg.dev/$PROJECT_
ID/$REPOSITORY_A/out-of-band@$DIGEST
        imagePullPolicy: Always
EOF

$ cat << EOF > cloudbuild.yaml
steps:
```

```
    - name: gcr.io/google.com/cloudsdktool/cloud-sdk
      id: Get kubeconfig and apply manifest of out of band app
      entrypoint: sh
      args:
        - '-c'
        - |
          gcloud container clusters get-credentials $CLUSTER_NAME
--region $REGION --project $PROJECT_ID
          kubectl apply -f k8s.yaml -n $NAMESPACE_01
serviceAccount: 'projects/${PROJECT_ID}/serviceAccounts/${GCP_
SA_NAME_01}@${PROJECT_ID}.iam.gserviceaccount.com'
options:
  workerPool: 'projects/$PROJECT_NUM/locations/$REGION/
workerPools/$PRIVATE_POOL_NAME'
  logging: CLOUD_LOGGING_ONLY
EOF
$ gcloud builds submit . --config=cloudbuild.yaml
```

When navigating to `out-of-band-app` in your GKE cluster, you should see an error similar to the one here:

Figure 8.14 – Error showing that out-of-band-app does not have the attestation required

Binary Authorization is flexible and extensible; you can also create policies based on custom attestations and metadata or take advantage of integrations with security scanning and create policies based on vulnerability levels associated with the contents of your artifacts.

Using Binary Authorization to create policies that adhere to your requirements and organizational policies is best practice—you can find additional examples of policies and use cases here: `https://cloud.google.com/container-analysis/docs/ods-cloudbuild#build_your_container`.

Summary

In this chapter, we reviewed how to set up a secure software delivery pipeline in Cloud Build for multiple teams sharing a GKE cluster. We set up secure infrastructure using private connectivity and per-build GCP service accounts with minimal permissions. We introduced human-in-the-loop review with Cloud Build approvals while ensuring that we only admitted trusted and verified container images into our GKE cluster.

These features and practices enable you to begin to implement a more secure software delivery pipeline to GKE, one of the most popular runtimes in Google Cloud.

Next, we will shift focus to using Cloud Build to automate workflows with a more developer-centric, serverless platform—Cloud Run.

9
Automating Serverless with Cloud Build

Serverless platforms often focus on developer experience, seeking to ensure developers focus on their code with as little operational overhead as possible. In Google Cloud, platforms such as Cloud Functions and Cloud Run have sought to provide seamless developer experiences that expose little operational work to deploy production-ready code.

That said, these platforms do not ignore the need for certain operational tasks, but provide simple and powerful abstractions that developers can use to manage their code in production. This can include tasks such as building code into an artifact, rolling out new versions of code, scaling code horizontally, connecting to services over a private network, and more.

Cloud Build is often used to orchestrate such activities. With native integrations into Google's serverless offerings, Cloud Build helps developers achieve these tasks while still remaining true to the vision of serverless – focus on code, not on operations.

In this chapter, we'll cover the following topics:

- Understanding Cloud Functions and Cloud Run
- Building containers without configuration
- Automating tasks for Cloud Run and Cloud Functions

Understanding Cloud Functions and Cloud Run

We will begin by reviewing the "serverless" platforms in scope for this chapter, Cloud Functions and Cloud Run. With an understanding of what developers are responsible for in each, we can then review how to use Cloud Build to manage those responsibilities.

Cloud Functions

When looking to run code with as little operational overhead as possible, **Cloud Functions** is the first place most developers start. On this platform, developers can choose a supported language runtime that Cloud Functions supports and write small chunks of code. Developers can then deploy that code to Cloud Functions, and it will run in production without developers needing to do the following:

- Build their code into an artifact
- Build a container to run their code
- Create infrastructure to run their code
- Manage infrastructure to run and scale their code
- Set up connectivity to instances of their code
- Configure logging and monitoring for their running code

Developers instead can focus on writing their code, deploying to Cloud Functions with a single `gcloud` command:

```
$ gcloud functions deploy function-name -source .
```

Cloud Functions supports a specific set of versions of runtimes from languages including **Python**, **Node.js**, **Go**, and **Java**. You can find the full list of currently supported versions here: `https://cloud.google.com/functions/docs/writing`.

The most common pattern used with Cloud Functions is writing functions that are event-driven, that is, they execute in response to the occurrence of a specific event. This could be a change to a database, the creation of a cloud resource, or a response to a webhook.

Cloud Run

Cloud Run aims to provide a developer-friendly platform for those who already have or will have a vested interest in running their code in containers, but still seek as little overhead as possible.

Once a developer has a container, they can deploy it to Cloud Run with a single `gcloud` command:

```
$ gcloud run deploy my-backend --image=us-docker.pkg.dev/
project-name/container-image
```

The desire or necessity to run code in containers can come from multiple requirements:

- Developers want to ensure consistent behavior across multiple execution environments, from development to production.

- Developers in an enterprise may need to be prepared to have their code easily migrate to a different platform, such as Kubernetes.

- Developers may be deploying the code to multiple locations in need of a unifying packaging standard for deploying that code.

For these developers, they may require the need to build their own containers first, and then bring them to a "serverless" platform in which operations are minimal. Cloud Run provides such a platform, allowing developers to use containers as the deployment mechanism while still taking advantage of Cloud Run managing the following tasks:

- Creating the infrastructure to run their code

- Managing infrastructure to run and scale their code

- Setting up connectivity to instances of their code

- Configuring logging and monitoring for their running code

Cloud Run also allows developers to make a distinction between "services" and "jobs." Cloud Run services allow for long-running services that are accessible via an HTTPS endpoint, while Cloud Run jobs allow for run-to-completion services that can run instances to perform work in parallel, assigning each instance an index that developers can use to distribute work.

Finally, Cloud Run allows developers to have services adhere to the Knative API.

Knative is an open source project that enables operators to host their own, self-managed serverless platform atop a Kubernetes cluster. Because Cloud Run is adherent to the Knative API specification, when users deploy a Cloud Run service, they also get a configuration that can be used to deploy the same container to a Kubernetes cluster running anywhere.

This makes Cloud Run not only a serverless container platform, but one that can extend to environments external to Google Cloud, requiring minimal to no configuration changes.

Cloud Functions 2nd gen

There is a newer version of Cloud Functions, called **Cloud Functions 2nd gen** (`https://cloud.google.com/functions/docs/2nd-gen/overview`), released in 2022. This will be the primary version of Cloud Functions that we will interact with in the remainder of this chapter. One of the critical changes with Cloud Functions 2nd gen is that it is built atop Cloud Run.

This means that Cloud Functions provides an abstraction layer atop Cloud Run that satisfies the requirement to build a container and deploy it to Cloud Run, allowing users not familiar with containers to still use the platform. It maintains the developer contract that the previous version of Cloud Functions established: write code and let the platform handle the rest.

Consistent with the first version, deploying Cloud Functions 2nd gen can also be done with a single command:

```
$ gcloud functions deploy function-name --gen2 -source .
```

There are multiple improvements for this re-platforming under the hood:

- Longer code execution timeouts

- Support for concurrent requests to a single instance of code

- Support for traffic splitting between revisions of code

- Support for larger instance types

Cloud Functions 2nd gen also integrates with Google Cloud's **Eventarc** (`https://cloud.google.com/eventarc/docs`), which is a managed event-delivery service with dozens of out-of-the-box event supports from Google Cloud resources.

Comparing Cloud Functions and Cloud Run

Because Cloud Functions 2nd gen is built as an abstraction layer atop Cloud Run, it should be noted that there can be an overlap in the use case. While a full discussion of when to use one versus the other is out of the scope of this chapter, it should be noted that generally, Cloud Functions should typically perform a single task, performing some sort of processing in response to an event. In addition, while Cloud Functions 2nd gen does use Cloud Run under the hood, you cannot customize the containers it creates and deploys. Rather, you are subject to the provided runtimes and working within those constraints.

Cloud Run, on the other hand, is a full-fledged container-as-a-service platform. This means that you can build anything into a container image and run it on Cloud Run – from web services to monolithic applications!

With that, let's review how developers can simplify the production of the standard package – the container – that Cloud Run and Cloud Functions 2nd gen use. This simplification can help minimize overhead cognitively, usage of personal workstation resources, and local optimizations that lead to silos.

Building containers without a build configuration

If the mission of building on serverless is to make running software as easy as possible for developers, it is important to consider areas where you can reduce cognitive overload – especially in areas that are not unique or specific to providing their business with unique value.

Dockerfile

For developers who have worked with containers already for their applications, they can bring their Dockerfile to Cloud Build, and *without* writing any build configuration, can run a build immediately:

```
$ gcloud builds submit --tag REGION-docker.pkg.dev/PROJECT_ID/
REPO/IMAGE_NAME
```

When using `gcloud`, it can detect that when you have a Dockerfile present but a build configuration missing, your intention is to submit a single-step build that runs Docker and builds your container image.

However, if you do not have prior history and experience with writing efficient and well-structured Dockerfiles, you may look to not only take advantage of not writing build configuration but also not optimizing a Dockerfile from scratch.

Language-specific tooling

Developers working heavily in a single language, such as Java or Go, may instead look to language-specific tooling that abstracts away the need for writing any Dockerfile or build, but rather a) handles that for you, while b) containing optimizations specific to your language and runtime for your app.

While the examples here are not exhaustive, let's review a few examples of language-specific tooling that helps developers build containers and are compatible with platforms such as Cloud Build.

For Java developers, tools such as **Maven** or **Gradle** are typically used to build projects into JAR files. Without language-specific tooling, developers would then need to step through the process of authoring a Dockerfile – that means ensuring that they choose the right base image, order their layers appropriately, and so on.

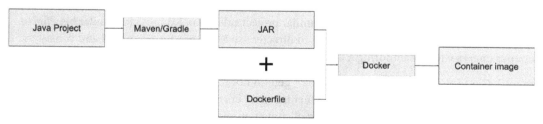

Figure 9.1 – Common workflow for Java developers without language-specific tooling

Google has open-sourced a project called **Jib**, which can build directly from a Java project and produce a container image without requiring developers to work through Dockerfiles. It has also been released as a plugin for both Maven and Gradle.

Figure 9.2 – Simplifying workflow for building containers using Jib

Java developers can then run Jib with Maven or Gradle as a build step within Cloud Build, offloading the build requirements from their own local resources. Once running builds in the cloud with Jib, they can also begin to take advantage of integrations with other Google Cloud services such as **Cloud Storage** for caching build files.

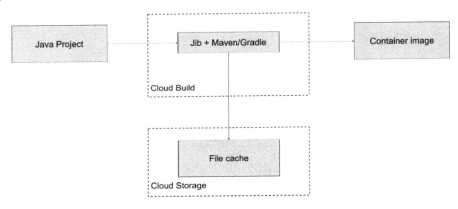

Figure 9.3 – Running Jib as a Cloud Build step and using Cloud Storage as a build cache

For Go developers, Google open-sources a similar project called **Ko**. Ko provides similar benefits to Jib, by providing those with Go applications to directly use Ko to produce standard and consistent container images without authoring a Dockerfile.

That said, Cloud Run and Cloud Functions 2nd gen do not natively expose the usage of language-specific tooling to build the underlying container images. Let's now review how Cloud Run and Cloud Functions 2nd gen abstracts away the building of container images: buildpacks.

Buildpacks

In Google Cloud, serverless offerings such as Cloud Functions and Cloud Run both make use of Google-supported **Cloud Native Buildpacks** (`https://buildpacks.io`) in combination with Cloud Build to deliver this experience to developers.

Cloud Functions and Cloud Run both aim to abstract away the process of using buildpacks to build container images, but it is still important to review their underlying fundamentals and how they interact with Google Cloud Build.

Cloud Native Buildpacks is an open source project that aims to provide developers with a streamlined way of building containers directly from source code without requiring developers to define the container themselves.

The way that it achieves this is by providing executables that inspect your source code and, based on that, will plan and execute the creation of your container image.

Buildpacks build container images using two primary phases:

- Detect
- Build

These phases occur in a resource called a **builder**. Google's builder is delivered as a container image hosted in **Container Registry**. You can find and inspect this container image by running the following command:

```
$ docker pull gcr.io/buildpacks/builder
```

Google's builder container image comes packaged with buildpacks that can build Go, Python, Java, and .NET applications. You can find the most current runtime support here (`https://github.com/GoogleCloudPlatform/buildpacks#general-builder-and-buildpacks`).

These buildpacks will run the detect phase on your source code using a detect executable. In the detect phase, the buildpack will determine whether your app has the prerequisites needed to build – in Python, we can think of the `requirements.txt` file; these prerequisites will vary based on your runtime.

If these prerequisites are met, then the build phase will commence. The build phase will build your application using a build base image, and then copy the output contents over to what is called the run base image to complete the creation of the image that hosts your app.

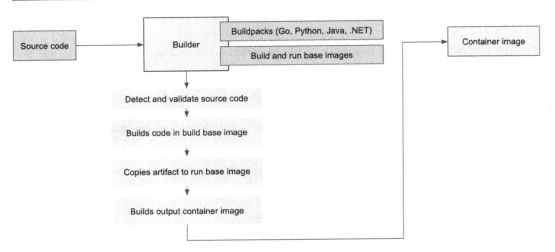

Figure 9.4 – Workflow for using a builder with buildpacks to build
a container image directly from the source

> **Note**
> Should you need to customize the builder or the run images used in this process, you can follow the process here: `https://github.com/GoogleCloudPlatform/buildpacks#extending-the-run-image`.
>
> This typically is required for users who need to install their own system-level packages into these images used.

This process is wrapped by a platform that abstracts away this process from the end user; one of the most common ways to consume this is via the `pack` CLI. This tool enables you to quickly run directly from your app directory and kick off the two phases to build your container image.

To use `pack` with Google's builder container image locally, you can run the following command from your application source code directory:

```
$ pack build my-app --builder gcr.io/buildpacks/builder
```

Running `pack` invokes the Docker daemon on the machine in which it is running, using this to execute the builder container. Many developers often face security constraints – and at times, resource constraints – that prevent them from being able to run Docker locally. Thus, Google Cloud Build provides native functionality for using a buildpack builder to build a container image on Cloud Build workers.

To use Cloud Build's native support for building container images using buildpacks, run the following command from your source code directory:

```
$ gcloud builds submit --pack image=gcr.io/$PROJECT_ID/my-app
```

Running this command will kick off a build in Cloud Build to build, tag, and push your container image to Container Registry.

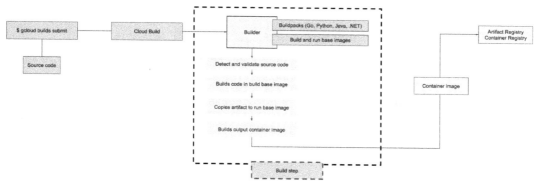

Figure 9.5 – Workflow for using a builder with buildpacks to build a
container image directly from the source using Cloud Build

This build will require access to the public internet to pull the builder, so keep in mind that if you are using a private pool, you must host the builder image yourself in your project's Container Registry or Artifact Registry.

For DevOps or security teams looking to establish standardized images with custom or bespoke tooling packaged into the final container image generated, you can also follow the same process.

Building and hosting your own builder allows you to provide your developers with standard but customized images by using the Google-provided builder image as a base image and then adding your own packages, similar to the following example:

```
# For demonstration purposes only, to run this build would
require setup beyond the scope of its inclusion
FROM gcr.io/buildpacks/builder:v1
USER root
RUN apt-get update && apt-get install -y --no-install-
recommends \
   subversion && \
   apt-get clean && \
```

```
    rm -rf /var/lib/apt/lists/*
USER cnb
```

Using Cloud Build, you can provide the custom builder image by setting it in the builder flag with submitting a build, similar to the following example:

```
$ gcloud builds submit --pack builder=gcr.io/$PROJECT_ID/
my-builder image=gcr.io/$PROJECT_ID/my-app
```

With this model, we get the best of both worlds:

- Developers can focus more on writing their code, as this method does not require any `Dockerfile` or Cloud Build configuration.

- Operators and security can establish best practices and standards, either using the ones out of the box from the buildpacks team or building atop these standards with their own bespoke tooling.

Now that we have reviewed the foundations of buildpacks, let's dive into how Cloud Run and Cloud Functions use Cloud Build to automate the use of buildpacks – amongst many other tasks.

Automating tasks for Cloud Run and Cloud Functions

In this section of the chapter, we will review numerous tasks related to Cloud Run and Cloud Functions that you can automate using Cloud Build.

Deploying services and jobs to Cloud Run

The first, and perhaps most simple task you can automate is doing a build and deployment of a container image. Earlier in this book, in *Chapter 4, Build Configuration and Schema* and *Chapter 7, Automating Deployment with Terraform and Cloud Build* we have reviewed multiple ways to use the Google Cloud APIs, namely the following:

- gcloud
- Terraform

Both wrap the APIs that allow developers to use the Google Cloud APIs to manage resources, including Cloud Run services and jobs.

From a developer experience perspective, when considering how to interact with Cloud Run from Cloud Build, it is preferred to use `gcloud` as it can be easier to reason about for developers looking to keep operational overhead to a minimum. Terraform is written in a JSON-like language called **HashiCorp Configuration Language** (**HCL**), which has its own syntax that developers must learn (`https://www.terraform.io/language/syntax/configuration`).

> **Note**
> It should be noted that many enterprises and organizations do require their developers to create all cloud resources using Terraform, for purposes of consistency and reusability.

First, let's review a simple build that mimics the build configurations we have reviewed earlier in the book, specifically in *Chapter 3, Getting Started – Which Build Information Is Available to Me?* and *Chapter 4, Build Configuration and Schema*:

```
# For demonstration purposes only, to run this build would
require setup beyond the scope of its inclusion
steps:
# Build the container image
- name: 'gcr.io/cloud-builders/docker'
  args: ['build', '-t', 'gcr.io/PROJECT_ID/IMAGE', '.']
# Push the container image to Container Registry
- name: 'gcr.io/cloud-builders/docker'
  args: ['push', 'gcr.io/PROJECT_ID/IMAGE']
# Deploy container image to Cloud Run
- name: 'gcr.io/google.com/cloudsdktool/cloud-sdk'
  entrypoint: gcloud
  args: ['run', 'deploy', 'SERVICE-NAME', '--image', 'gcr.io/
PROJECT_ID/IMAGE', '--region', 'REGION']
images:
- gcr.io/PROJECT_ID/IMAGE
```

The core difference here is the third build step, in which `gcloud` is utilized to deploy the container image to Cloud Run. While this is fairly straightforward, it does provide a useful opportunity to call attention to the specific builder being used, the `cloud-sdk` image.

In *Chapter 4, Build Configuration and Schema* we discussed the concept of Cloud Builders, that is, container images made available for build steps by the Cloud Build team. In the Cloud Builders project, there is already an image for `gcloud`. However, in this example, we use the `cloud-sdk` image provided by the official Cloud SDK team.

You can use a very similar build to kick off a Cloud Run job:

```
# For demonstration purposes only, to run this build would
require setup beyond the scope of its inclusion
steps:
# Build the container image
```

```
- name: 'gcr.io/cloud-builders/docker'
  args: ['build', '-t', 'gcr.io/PROJECT_ID/IMAGE', '.']
# Push the container image to Container Registry
- name: 'gcr.io/cloud-builders/docker'
  args: ['push', 'gcr.io/PROJECT_ID/IMAGE']
# Deploy container image to Cloud Run
- name: 'gcr.io/google.com/cloudsdktool/cloud-sdk'
  entrypoint: gcloud
  args: ['beta', 'run, 'jobs', 'create', 'JOB-NAME', '--image',
'gcr.io/PROJECT_ID/IMAGE', '--region', 'REGION']
images:
- gcr.io/PROJECT_ID/IMAGE
```

While it's useful to know that the `gcloud` build steps will suffice for deploying both Cloud Run services and jobs, Cloud Run has several life cycle features built into its resources. These features either natively use Cloud Build under the hood or can be automated using builds similar to the ones shown previously.

Deploying to Cloud Functions

For Cloud Functions, you can use the same `cloud-sdk` image to use `gcloud` to invoke the `gcloud functions deploy` command:

```
# For demonstration purposes only, to run this build would
require setup beyond the scope of its inclusion
steps:
- name: 'gcr.io/google.com/cloudsdktool/cloud-sdk'
  args:
  - 'gcloud'
  - 'functions'
  - 'deploy'
  - 'FUNCTION_NAME'
  - '--gen2'
  - '--region=FUNCTION_REGION'
  - '--source=.'
  - '--trigger-http'
  - '--runtime=RUNTIME'
```

When deploying code using the previous command, Cloud Functions actually "compiles" down to two things under the hood:

- Cloud Functions kicks off a single-step build in Cloud Build that runs the pack CLI on a default worker in Cloud Build.

- Cloud Functions takes the image built and automates the deployment of Cloud Functions and underlying Cloud Run resources.

> **Note**
>
> You can also specify a specific worker pool in Cloud Build to build the backing container image for your function code using the `--build-worker-pool` flag when using `gcloud` functions deploy.

If you are working with a net-new workload, it is recommended you adopt Cloud Functions 2nd gen over the first generation of Cloud Functions.

Going from source code directly to containers running in Cloud Run

We have now discussed how buildpacks work along with how you can invoke the use of buildpacks to build a container image using the Cloud Build API. We have also reviewed how Cloud Functions has natively baked in buildpacks into its own deployment method.

Cloud Run also exposes the usage of buildpacks in its resources, allowing developers to work directly with those APIs not only to run their code but also to build their containers automatically without any `Dockerfile` or YAML configuration.

The following example command displays one additional flag referencing where your source directory is:

```
$ gcloud run deploy my-app --source .
```

Similar to Cloud Functions 2nd gen, this command with the `--source` flag similarly "compiles" down to two `gcloud` commands:

- Creating and executing a build that runs the `pack` CLI on a default worker in Cloud Build to build a container image and push it to Container Registry

- Running `gcloud run deploy` with the image field set to be the container image that the previous step just built

This allows developers to quickly test out their application running packaged in a container image in Cloud Run.

> **Note**
> At publication time, unlike Cloud Functions 2nd gen, all builds kicked off through Cloud Run will schedule and execute on a worker in the default pool.

However, as we have discussed in previous chapters, more likely than not, you'll be implementing builds with triggers to automate the execution of the aforementioned steps.

To configure this within the context of Cloud Run, you will need to look at the Cloud Run UI console when creating a service – selecting **Continuously deploy new revisions from a source repository** and then **SET UP WITH CLOUD BUILD**:

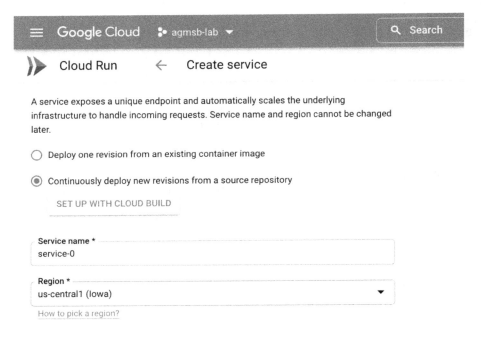

Figure 9.6 – Creating a service that is directly deployed from a connected source repository

Then, you can define the source code repository you want to utilize, define what branch or branches you want to trigger a fresh build off of using regular expressions, and select Google Cloud Buildpacks as the mechanism you want Cloud Build to use when building your container image.

> **Note**
> In order to complete this step, you will need to connect source repositories such as GitHub, GitHub Enterprise, or Bitbucket to Cloud Build. You can find the latest instructions on setting this up in the Cloud Build documentation (`https://cloud.google.com/build/docs/`).

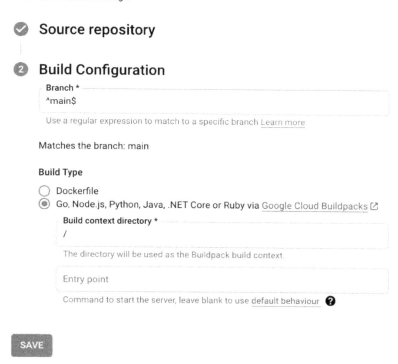

Figure 9.7 – Enabling builds with buildpacks for a Cloud Run service in the Cloud console

With this setup, changes to the specified branches in your source code repository will now trigger the same build set up by Cloud Run. While you do not have to, if you want to view the build or build logs, you can navigate to the Cloud Build console.

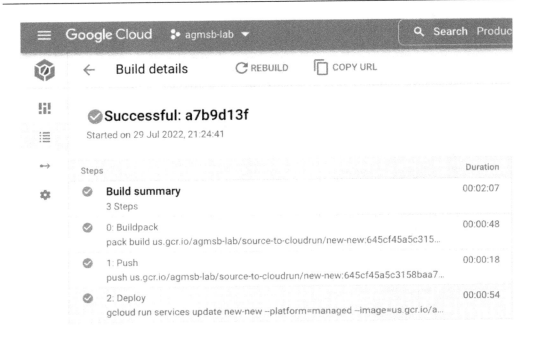

Figure 9.8 – The three-step build auto-created in Cloud Build by Cloud Run

With that said, Cloud Run will also surface build history for the overall service and build logs per revision directly in the Cloud Run console as well.

Figure 9.9 – Build history available in the Cloud Run console

Main container (Port 8080, 1 CPU, 512MiB memory)   ⌃

Image URL	us.gcr.io/agmsb-lab/source-to-cloudrun/new-new@sh... ⎗
Port	8080
Build	Cloud Build (logs)
Source	https://github.com/agmsb/source-to-cloudrun/comm... ⎗
Command and arguments	(container entrypoint)
CPU limit	1
Memory limit	512MiB

Figure 9.10 – Build logs available in the Cloud Run console

Cloud Run uses Cloud Build under the hood of its turnkey source to prod functionality. By natively weaving Cloud Build's functionality into Cloud Run resources, developers are able to achieve the ease of use that serverless platforms promise.

With that said, the method just reviewed provides an opinionated deployment model that automates the step-through of blue-green deployments.

A blue-green deployment is a method in which two versions of an app are deployed, and 1-100% of traffic is cutover from the older "blue" version to the newer "green" version:

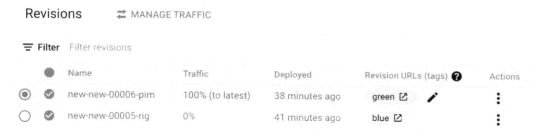

One of the highlights that Cloud Run brings to developers, however, is support for more complex rollout patterns in a declarative fashion.

Let's now discuss how you could use Cloud Build to automate these rollout patterns that are built into the Cloud Run API.

Progressive rollouts for revisions of a Cloud Run service

Cloud Run enables developers to automate traffic shaping between various revisions of their containers directly from its API. When using `gcloud`, you can use sets of specific flags to control shifting traffic amongst revisions.

When deploying a new revision of a Cloud Run service using `gcloud run deploy`, you can specify the below flag:

- `--no-traffic`

    This ensures that the revision you just deployed will not receive any traffic upon deployment.

Once deployed, you can then run the `gcloud run services update-traffic` command to granularly control the service:

- `--to-revisions`

    Using this flag allows you to explicitly specify percentages of traffic to named revisions of your service.

- `--to-latest`

    Using this flag will shift all traffic to the latest revision of your service.

While you could run these commands manually as well, let's review how you could integrate this functionality with source code management and triggers in Cloud Build.

Using the `--no-traffic` flag specifies that your new revision will be available and associated with your Cloud Run service, but by default, will have no traffic sent to it.

The following sample build step uses this flag:

```
# For demonstration purposes only, to run this build would
require setup beyond the scope of its inclusion
steps:
- id: Rollout version with no traffic
  name: 'gcr.io/google.com/cloudsdktool/cloud-sdk'
  args:
  - 'gcloud'
  - 'run'
  - 'deploy'
  - 'SERVICE_NAME'
  - '--image'
  - 'gcr.io/PROJECT_ID/IMAGE_NAME:IMAGE_TAG'
  - '--region'
```

```
    - 'REGION'
    - '--platform'
    - 'managed'
    - '--no-traffic'
```

This effectively allows for a blue-green deployment model in which you are in control of when the cutover happens – as both revisions are associated with the Cloud Run service at the same time.

This has the added benefit of not incurring any additional costs that blue-green deployment models typically would in a VM or Kubernetes cluster, as Cloud Run only charges by the compute minute – if our new revision serves no traffic, it costs nothing.

With the new revision now deployed, you can begin with implementing a progressive rollout of traffic. The `--to-revisions` flag allows you to explicitly name the revision or specify LATEST and define a percentage of traffic to be directed to that named revision.

The following sample build step uses this flag:

```
# For demonstration purposes only, to run this build would
require setup beyond the scope of its inclusion
steps:
- id: Send 20% of traffic to latest version
  name: 'gcr.io/google.com/cloudsdktool/cloud-sdk'
  args:
  - 'gcloud'
  - 'run'
  - 'services'
  - 'update-traffic'
  - 'SERVICE_NAME'
  - '--to-revisions'
  - 'LATEST=20'
  - '--region'
  - 'us-central1'
```

Putting this into practice is considered a method of canary deployments, in which you send a subset of production traffic to your new or "canary" revision. In this case, 20% of production traffic.

During this time that your canary and your previous or "stable" co-exist, typically, you will use this time to perform canary analysis. During canary analysis, you can compare the canary revision to the stable revision, observing metrics such as latency to make a judgment as to whether or not the canary revision can be promoted to receive all traffic.

When you are ready to make this judgment, the `--to-latest` flag allows you to explicitly send all production traffic to your latest revision, effectively promoting that revision to production.

The following sample build step uses this flag:

```
# For demonstration purposes only, to run this build would
require setup beyond the scope of its inclusion
steps:
- id: Send 100% of traffic to latest version
  name: 'gcr.io/google.com/cloudsdktool/cloud-sdk'
  args:
  - 'gcloud'
  - 'run'
  - 'services'
  - 'update-traffic'
  - 'SERVICE_NAME'
  - '--to-latest
  - '--region'
  - 'us-central1'
```

This model also expands to Cloud Functions 2nd gen. While Cloud Functions 2nd gen abstracts away much of what users need to do with Cloud Run, traffic management still goes through the Cloud Run API.

Using Cloud Build to orchestrate this can enable you to trigger these respective processes manually, or in accordance with changes to your source code repository.

Traffic management is not the only complex deployment pattern that can be utilized with Cloud Run and Cloud Build. Let's now review how you can implement security best practices when deploying to Cloud Run.

Securing production with Binary Authorization

Given that Cloud Run allows developers, in theory, to deploy any container image they would like, how can administrators prevent developers from deploying arbitrary images, or worse, prevent bad actors from deploying harmful or malicious images?

Similar to what we reviewed for GKE in *Chapter 8, Securing Software Delivery to GKE with Cloud Build* **Binary Authorization** is also a feature that is available to Cloud Run developers.

This is the functionality that performs admission control for container images upon deployment, denying container images that violate predefined policies by cloud or security administrators.

While it is possible to enable Binary Authorization on a Cloud Run service level, it is recommended to enforce that Cloud Run services have Binary Authorization enabled at a higher level, such as an Organization, a Folder, or a Project.

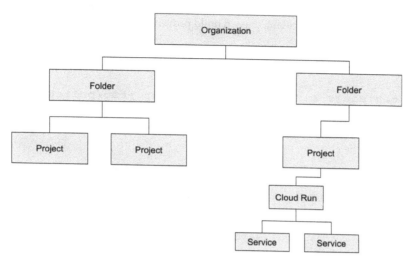

Figure 9.11 – Hierarchy of a Cloud Run service resource

If it is to be set only on a per-service basis, any developer with Cloud Run permissions can just as easily remove Binary Authorization from their service.

You can enable this at an Organization, Folder, or Project level with the following gcloud command, adding the respective resource ID to the --organization, --folder, or --project flag:

```
$ gcloud resource-manager org-policies allow run.
allowedBinaryAuthorizationPolicies default
```

With this in place, you now can create a Cloud Run service using gcloud run deploy while passing the --binary-authorization=default flag.

Similar to the example laid out in *Chapter 8, Securing Software Delivery to GKE with Cloud Build* you can now update your admission policies by running the following commands:

```
$ gcloud container binauthz policy export > /tmp/policy.yaml
```

This grabs the default admission policy with Binary Authorization enabled. You can edit this policy file by adding a block like the following into the policy file under the defaultAdmissionRule:

```
...

requireAttestationsBy:
```

```
        - projects/PROJECT_ID/attestors/built-by-cloud-build
    ...
```

This is the same attestation included in the previous chapter, which will only admit images with an attestation signed by Cloud Build, proving trusted build provenance for the container images running in Cloud Run.

With the policy file updated, you can now re-import the policy into Binary Authorization:

```
$ gcloud container binauthz policy import /tmp/policy.yaml
```

When violating policies at deployment time, Cloud Run will now return error messages similar to the following screenshot:

Figure 9.12 – Violating Binary Authorization policies at deployment time

You can also create your own attestor to sign attestations to your container images and have Binary Authorization enforce those by following this same process!

Summary

In this chapter, we reviewed how Cloud Run and Cloud Functions work in Google Cloud, as well as how they natively integrate Cloud Build into making serverless deployment simpler and empowering developers to focus on their code as much as possible.

We also reviewed how developers can use Cloud Build to automate and extend the native functionality provided in the Cloud Run and Cloud Functions APIs.

With a deeper understanding of how Cloud Build can work with infrastructure as code, Kubernetes clusters, and serverless platforms, we will now move on to Cloud Build practices that can help you run builds at scale and in production for any use case.

10
Running Operations for Cloud Build in Production

So far in this book, we've described the value of managed services, how to use Cloud Build, described Cloud Build capabilities, and provided end-to-end examples of Cloud Build in action. In this chapter, we are going to highlight specific Cloud Build capabilities and introduce new concepts to help you leverage Cloud Build in a production environment. The topics in this chapter will help you optimize your use of Cloud Build and the observability of your builds.

In this chapter, we will cover the following topics:

- Executing in production
- Configurations to consider in production
- Speeding up your builds

Executing in production

As a managed service, Cloud Build is leveraged by specifying build steps that result in a pipeline, whether it's building code or manipulating infrastructure components. It does not distinguish between environments; a pipeline that is running in your development environment is treated the same as a pipeline running in your production environment. They both will have the same Cloud Build **Service-Level Agreement** (**SLA**) of 99.95% (`https://cloud.google.com/build/sla`). The differences between environments can be, but are not limited to, the following:

- The uniqueness of the build steps in your configuration
- Different compute resources (default or private pools) defined in your project
- Security boundaries defined (identity, role bindings, permissions, and network resources)

For these criteria referenced, it is important to decide on the project(s) that will be hosting your build pipelines. This could be a factor of what your build pipelines are responsible for, and the roles of members within the organization that are responsible for interacting with the pipelines.

Organizations decide on the pattern that works best for them; however, the most common pattern is to host production separately from resources that are considered non-production. Pipelines, however, tend to blur the lines between non-production and production resources. For instance, there may be a neutral set of projects just for pipelines. When an integration pipeline is responsible for generating a container image to be used by downstream environments, it must have access to a shared container registry (for example, an **Artifact Registry**) that it can push built images to. Once the images are built the first time, it is a best practice that the same image is used in the remainder of the downstream environments from development to test to production. If it needs to communicate with other tools that deploy into other environments, it will need permission to communicate with the tool as well. The image may also be signed with specific attestations that it has passed through each of the pipelines for each respective environment.

Once the type of project has determined where the Cloud Build service pipelines will be executed, it is important to ensure that sufficient resources can be allocated. This requires you to understand the quotas and limits of the Cloud Build service (`https://cloud.google.com/build/quotas`). At the time of writing, the following are quotas for the Cloud Build service. Please refer to the aforementioned link for the most up-to-date information:

Resource	Description	Default Limit	Can Be Increased	Scope
Private pool	Number of private pools	2 to 10	Yes.	Per region
Build triggers	Number of build triggers	300	Yes.	Per region
CPU	Number of concurrent CPUs run in a regional private pool	0 * to 2,400	Yes. When this quota is filled, requests for additional CPUs are queued and processed serially.	Per region
CPU	Number of concurrent CPUs run in a regional default pool	5 to 100	No. If you require more than 100 concurrent CPUs, use `private pools`.	Per region

Request	Number of concurrent builds run in a global default pool	10 to 30	No. If you want to run more than 30 concurrent builds, consider using `private pools`.	

Table 10.1 – Quotas for the Cloud Build service

An important factor to bear in mind from the preceding table is how many concurrent builds you will need to run in each respective project. For instance, the project where development and integration pipelines run may be more active than pipelines required for production environments or vice versa, depending on your environment. This emphasizes the importance and flexibility of leveraging private pools for your pipelines. Private pools allow the following:

- You can specify more machine types for your builds
- You can have more concurrent builds – that is, 30+
- You can gain access to VPC-connected resources
- You can have network flexibility

Determining how to organize Cloud Build pipelines, capabilities to leverage, and identifying dependent resources are important considerations in the planning process.

Leveraging Cloud Build services from different projects

If the goal is to leverage Cloud Build services across multiple teams or environments, sharing the same Google Cloud project will require additional considerations. In *Chapter 6, Managing Environment Security*, we discussed the principle of least privilege, where we defined specific Cloud Build user-specified service accounts for each configured trigger. When teams share the project, you may have service accounts that are hosted in a different project. If that is the case, additional permissions might be needed for the service account to be used in the Cloud Build project where the pipeline will be executing.

The following are for example and illustrative purposes, as you may have a different repository configured in your environment.

Set up the appropriate variables to be utilized in the remainder of the commands for this example as your values will differ based on your environment. The GCS bucket should be unique:

```
$ SA_NAME=build-dev
$ DEV_PROJECT=dev-project
```

```
$ BUILD_PROJECT=build-project
$ ARTIFACT_REGISTRY_IMAGE_REPO=image-repo
$ GCS_BUCKET=packt-cloudbuild-dev-bucket
$ GIT_REPO=packt-cloudbuild
$ GIT_OWNER=user
```

> **Note**
> In the following example, `dev-project` must have the `iam.disableCrossProjectServiceAccountUsage` organizational policy disabled. Organizational policies are configured at the organizational level but can be applied to folders and individual projects, depending on the type of constraint. Constraints can vary but determine what can be done, beyond what IAM permissions define, which determine who can do what.

The following diagram shows what this example will create, as well as the relationship between a build and a user-specified service account:

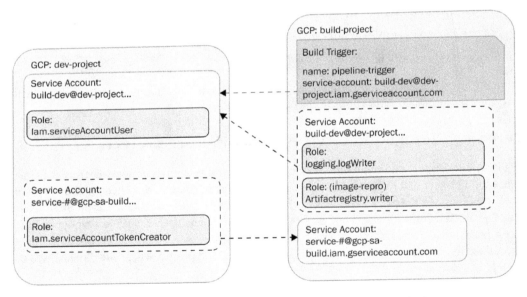

Figure 10.1 – User-specified service account in a separate project

The preceding diagram shows the relationship between the two projects and the permissions associated with each project to leverage a user-specified service account in another project.

Create the service account to be used as your user-specified service account for the Cloud Build pipeline:

```
$ gcloud iam service-accounts create ${SA_NAME} \
    --description="Build Dev Service Account" \
    --project ${DEV_PROJECT}
```

The service account needs Service Account User permissions so that it can be used by the Cloud Build service agent in the project that will use the Cloud Build service:

```
$ gcloud projects add-iam-policy-binding ${DEV_PROJECT} \
  --member=serviceAccount:${SA_NAME}@${DEV_PROJECT}.iam.
gserviceaccount.com \
  --role=roles/iam.serviceAccountUser
```

The project number for the Google Cloud Project that will leverage the Cloud Build service is required so that service account impersonation can be performed. We set the permission for the Cloud Build service agent to be able to create service account tokens to execute commands on behalf of the service account hosted on our development project:

```
$ export BUILD_PROJECT_NUMBER=$(gcloud projects list \
  --filter="${BUILD_PROJECT}" \
  --format="value(PROJECT_NUMBER)"
)
$ gcloud projects add-iam-policy-binding ${DEV_PROJECT} \
    --member="serviceAccount:service-${BUILD_PROJECT_NUMBER}@
gcp-sa-cloudbuild.iam.gserviceaccount.com" \
    --role="roles/iam.serviceAccountTokenCreator"
```

Although the service account is located on the development project, we give the service account permissions to write logs into the build project where the Cloud Build service will be utilized:

```
$ gcloud projects add-iam-policy-binding ${BUILD_PROJECT} \
  --member=serviceAccount:${SA_NAME}@${DEV_PROJECT}.iam.
gserviceaccount.com \
  --role=roles/logging.logWriter
```

Since we are using a user-specified service account, we will need to configure with our `cloudbuild.yaml` configuration to only log. The following code shows part of the `cloudbuild.yaml` file. Later in this chapter, we will walk through a scenario where we store the logs in a **Google Cloud Storage** (**GCS**) bucket:

```
...
options:
  logging: CLOUD_LOGGING_ONLY
...
```

The trigger is created on the build project, but the user-specified service account is the one we created in our development project:

```
$ gcloud beta builds triggers create github \
    --repo-name=${GIT_REPO} \
    --repo-owner=${GIT_OWNER} \
    --branch-pattern=^main$ \
    --name=pipeline-trigger \
    --build-config=cloudbuild.yaml \
    --project=${BUILD_PROJECT} \
    --service-account=projects/${DEV_PROJECT}/
serviceAccounts/${SA_NAME}@${DEV_PROJECT}.iam.gserviceaccount.
com
```

In this example, we are pushing a container image to Artifact Registry's image repository, which is also hosted on the same Cloud Build project. Permissions will be required to write to the image repository:

```
$ gcloud artifacts repositories add-iam-policy-binding
${ARTIFACT_REGISTRY_IMAGE_REPO} \
    --location=us-central1 \
    --member=serviceAccount:${SA_NAME}@${DEV_PROJECT}.iam.
gserviceaccount.com \
    --role=roles/artifactregistry.writer \
    --project ${BUILD_PROJECT}
```

This is the sample output for the build details of the triggered build from the preceding example. The Project ID that the build was executed from is in `build-project` and the service account is from `dev-project`. We can get this information by describing the build:

```
$ gcloud builds describe 2ddd7e2e-1b75-49e2-b3a9-4781ad509f18
--project build-project
```

The following is the output of running the preceding command, which identifies the user-specified service account being utilized:

```
id: 2ddd7e2e-1b75-49e2-b3a9-4781ad509f18
...
projectId: build-project
...
serviceAccount: projects/dev-project/serviceAccounts/build-dev@
dev-project.iam.gserviceaccount.com
```

Since the build is running in the build project and it automatically triggers when there is a commit to the repository, it may very well be the scenario that developers or specific team members only have access to specific projects – in this case, `dev-project`. In the previous example, we wrote logs into `build-project`, but we could set up a sink and forward logs (`https://cloud.google.com/logging/docs/export/configure_export_v2#supported-destinations`) from the build-project to `dev-project`.

Another approach consists of writing to a GCS bucket that is managed and hosted in `dev-project`. The build log data is created by `build-project` and is stored in `dev-project` for further analysis:

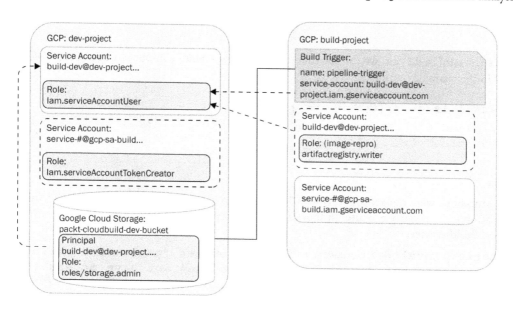

Figure 10.2 – Logging to a GCS bucket in another project

The preceding diagram shows how the build writes to the GCS bucket, with the specific permissions required by the user-specified service account at the bucket level.

First, let's create the bucket in `dev-project`:

```
$ gsutil mb -p ${DEV_PROJECT} \
    gs://${GCS_BUCKET}
```

To allow the Cloud Build service to use our user-specified service account to write to the GCS bucket, we need to give the necessary permissions to our service account:

```
$ gsutil iam ch \
    serviceAccount:${SA_NAME}@ ${DEV_PROJECT}.iam.
gserviceaccount.com:roles/storage.admin \
    gs://{GCS_BUCKET}
```

We will also need to tell Cloud Build that we want to store logs in a GCS bucket through our `cloudbuild.yaml` file:

```
...
logsBucket: "gs://${REPLACE_WITH_YOUR_GCS_BUCKET_NAME}"
options:
  logging: GCS_ONLY
...
```

After a successful build run, we can see that the log is now available in our storage bucket. We can see this in the example output:

```
$ gsutil ls gs://{GCS_BUCKET}
gs://packt-cloudbuild-dev-bucket/log-44045da7-3a03-4f9d-8aa0-
f9599e3efd47.txt
```

> **Note**
>
> If permissions were not set properly for Cloud Build to log to the GCS bucket, the build will not be able to properly execute. In that case, you may receive the following error from the Cloud Build service:
>
> ```
> generic::invalid_argument: invalid bucket "packt-cloudbuild-
> dev-bucket"; builder service account does not have access to
> the bucket
> ```

When leveraging different projects, storing log data in different projects may be necessary. The example in this section shows how this can be done.

Securing build triggers even further

In *Chapter 6*, *Managing Environment Security,* we discussed **Virtual Private Cloud Service Controls (VPC-SC)** as a mechanism to help prevent data exfiltration for GCP services, including Cloud Build. Another mechanism available in Cloud Build to prevent triggers from external sources to secure your build environment is through organizational policies. With this constraint set, we can specify which domains can perform a function or deny specific domains.

So far, our examples have been triggered through `github.com`. Now, we will deny a trigger coming from `github.com`. To define this, we must set a constraint and specify permissions:

```
name: projects/**REDACTED_project_number**/policies/cloudbuild.
allowedIntegrations
spec:
  inheritFromParent: false
  rules:
    - values:
        deniedValues:
          github.com
```

We are going to deny triggers that are sourced from `github.com`. Notice the constraint that we will be enforcing – that is, `cloudbuild.allowedIntegrations`. Once enforced, the trigger from the previous section will no longer be able to execute and create a build. The following Cloud Logging output snippet shows the reason why the build is unable to trigger and execute:

```
status: {
code: 9
message: "Constraint "constraints/cloudbuild.
allowedIntegrations" violated for "**REDACTED_project_id**"
attempting to process an event from "github.com""
}
```

This provides another capability to help secure your builds.

Notifications

When you initially set up builds or manually trigger builds, you may be interacting with and watching the builds as you incrementally test out the build configuration. You would be aware of when builds are failing or succeeding. However, this is not a typical usage pattern when builds are automatically triggered based on external events such as a commit to a **Source Control Management (SCM)** repository. Organizations have different types of mechanisms to notify developers and operations teams of the status of builds. These can vary from systems that post messages regarding the build status in a chat infrastructure, email notifications, or custom infrastructure to allow teams to digest build event status.

Depending on the verbosity required, organizations may end up tracking the following:

- Git commits into a repository branch

- Merges between branches

- Triggered build pipeline status (for example, successful, failed, pending, and so on)

- The build success rate

This helps give organizations insight into the health and frequency of their builds so that they can be used to optimize or identify bottlenecks within an organization.

Cloud Build uses Cloud Pub/Sub (`https://cloud.google.com/pubsub`) by publishing messages on a specific topic. While these messages can be consumed by subscribing to the topic using Cloud Pub/Sub clients, Cloud Build includes a Cloud Build notifier (`https://cloud.google.com/build/docs/configuring-notifications/notifiers`) client that can also subscribe to the topic, consume messages, and push messages to supported platforms. The Cloud Build notifier clients are container-based images that run on Cloud Run (`https://cloud.google.com/run`), which is a managed serverless platform that scales up and down based on usage. The following table shows the available notifier-supported platforms:

Notifier	Description
bigquery	Writes build data to a BigQuery table
googlechat	Uses a Google Chat webhook to post messages to a Google Chat space
http	Sends a JSON payload to another HTTP endpoint
slack	Uses a Slack webhook to post messages to a Slack channel
smtp	Sends emails via an SMTP server

Table 10.2 – Cloud Build notifier platforms

The preceding table shows the available notifiers at the time of writing. If you are using a platform that is not listed here, you can develop a notifier using the example provided (`https://cloud.google.com/build/docs/configuring-notifications/create-notifier`).

Deriving more value from logs

Logs provide important details for debugging and validating the progress of builds. The output of logs typically requires an individual to view the logs to make sense of the output. When aggregating builds that have numerous pipeline runs, viewing logs may not be viable. In the previous section, we discussed notifications as a way to aggregate build pipeline event statuses that are outputted to understand the health of pipelines. What if we wanted to know how many times during a build step certain log entries are being outputted? We can use this to identify certain patterns and help with automation or approvals. We can also alert if the same log output accumulates to a specified threshold.

Logs-based metrics can be achieved through a Cloud Logging capability. Suppose, for instance, while building a Node.js application, we want to track how many times we get deprecation warnings. In this example, the Node.js bookshelf application (`https://github.com/GoogleCloudPlatform/nodejs-getting-started/tree/main/bookshelf`) is being built and outputs the following Cloud Build log output:

```
...
Step 4/6 : RUN npm install
Running in 30f414a77c43
npm WARN deprecated ...
...
```

We want to be able to track how many times this is outputted during our builds. Given this is a Node.js warning, the builds will still succeed, but we want to see how many times this is occurring; if it exceeds a certain threshold, we may want to take action. Builds may be succeeding now, but for how long before they start failing?

Let's create a log-based counter metric for this output:

```
$ gcloud logging metrics create npm_dep_warning \
  --description "NPM Deprecation Warnings" \
  --log-filter "resource.type=build AND npm WARN deprecated"
```

The counter metric increments by 1 each time the evaluation of the log filter returns a new entry. The count can be viewed through the Cloud Monitoring **Metrics Explorer** pane. More details can be found in the documentation (`https://cloud.google.com/monitoring/charts/metrics-selector`).

The following screenshot shows that within 10 minutes, the log-based metric has counted that our `npm WARN deprecated` log entry has occurred twice:

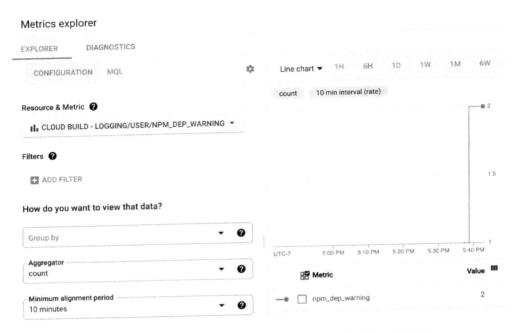

Figure 10.3 – Cloud Monitoring Metrics Explorer NMP_DEP_WARNING

We can alert based on any threshold that makes the most sense. In the following example, we have an alerting policy that triggers if it happens five times within a rolling 10-minute window. More details on how to create the alert can be found in the documentation, which can be created through the **user interface (UI)** or the API (`https://cloud.google.com/monitoring/alerts/using-alerting-ui`).

The following example alert policy will create an incident when the warning threshold goes beyond the count of 5:

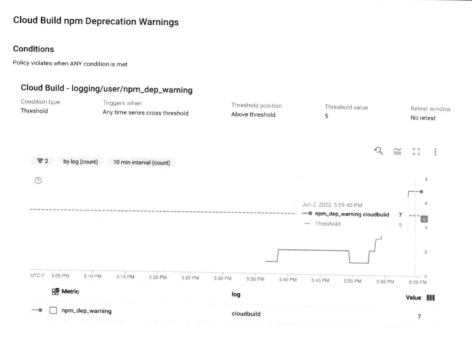

Figure 10.4 – Cloud Monitoring alert policy Cloud Build npm deprecation warnings

From the **Alerting** pane, we can see that an incident has been raised and what action may need to be taken. Cloud Monitoring notification channels include a variety of pre-built integrations for tools such as Cloud Mobile App, Cloud Pub/Sub, Email, SMS, SMS, PagerDuty, Slack, and webhooks:

Figure 10.5 – Cloud Monitoring alert incident

While this example consisted of tracking how many times an entry showed up in a log, we can also track the inverse – that is, how many times something did not show up during a specified period.

Log-based metrics also support another metric known as the distribution metric, which can measure the latency or duration between entries to identify bottlenecks within builds. The combination of counter and distribution metrics provided by Cloud Logging and Cloud Monitoring can help extrapolate more details beyond default build statistics.

Executing in production requires additional considerations beyond just running a pipeline. It is important to test and identify the best machine type for each build that results in acceptable build times and associated costs. Private pools provide a larger selection of machine types, while also offering additional network and security capabilities. Depending on your organization's Google Cloud project architecture, team roles, and personas, access to different projects may need to be considered. Getting the most out of your logs and deriving patterns from logs and notifications are essential to Cloud Build running in the background and running smoothly.

Configurations to consider in production

In *Chapter 4, Build Configuration and Schema,* we covered a few configuration options that can be used to customize how Cloud Build will execute your build. In production, there are a few configurations that stand out and should be configured to optimize your builds in production.

We discussed Cloud Build request limits, which depend on the type of pool that is selected, in the previous section. Regardless of the type of pool, we can prioritize or determine that certain build pipelines should fail if they are queued for too long.

To configure this setting, we will use the `queueTtl` syntax in our build configuration. Once a build is in the queue, the timer starts. If it is unable to be executed before the duration is specified in the `queueTtl` configuration, it will be canceled, removed from the queue, and given an `EXPIRED` status. This can help provide a level of prioritization to the builds. If there are too many in the queue, the builds will back up, and critical tasks may be delayed, which can impact productivity. One way around this could be to increase the amount of compute resources in a private pool or switch from the default to a private pool. Organizations may have defined budgets or have certain builds that can be requeued at a later time when builds may be less busy.

Making builds more dynamic

Variables can be used in Cloud Build's configuration to make pipelines more dynamic and reusable. In *Chapter 5, Triggering Builds,* we covered using variables to perform a Git checkout with the commit SHA and storing a Git SSH key from a secret. The commit SHA value is not known during configuration time and could change for each triggered build. The SSH key should preferably be managed outside so that it is stored on a platform that has more secure secrets.

Cloud Build includes a few default metadata-related variables that can be used in the build, such as `${PROJECT_ID}` and `${BRANCH_NAME}` to trigger the Git repository. Default variables provide metadata, but user-defined variables can allow for additional metadata, as well as drive behaviors within a build pipeline. You may want to leverage a variable to define the environment for the build, such as `${_ENV}`. This can be used within your build configuration to set particular metadata as part of the build or runtime configuration.

To make your build steps or pipelines more flexible, you can use variables within variables. Cloud Build enables this capability through the use of a configuration option known as `dynamic_substitutions`.

Dynamic substitutions are automatically enabled on build triggers, but builds that are manually invoked require this setting to be set to `true` in the build configuration.

Let's walk through an example of a Docker build. Here, we want to use user-defined variables to customize the container image repository, name, and tags. We also want to use dynamic substitutions to build out the full image URL.

We will continue to use the Node.js bookshelf application (`https://github.com/ GoogleCloudPlatform/nodejs-getting-started/tree/main/bookshelf`) for this example. The following code is from the `cloudbuild.yaml` configuration file, which builds a container image:

```
steps:
- name: gcr.io/cloud-builders/docker
  args:
  - "build"
  - "-t"
  - "${_IMAGE_URL}:${_BUILD_ID}"
  - "-t"
  - "${_IMAGE_URL}:${_TAG_VERSION}"
  - "."
substitutions:
  _IMAGE_URL: "${_IMAGE_BASE}/${_IMAGE_NAME}"
  _IMAGE_BASE: "us-central1-docker.pkg.dev/${PROJECT_ID}/image-repo"
  _IMAGE_NAME: "nodejs-bookshelf"
  _TAG_VERSION: "dev"
options:
  dynamic_substitutions: true
images:
- ${_IMAGE_URL}:${_BUILD_ID}
- ${_IMAGE_URL}:${_TAG_VERSION}
```

The dynamic substitution setting has been configured because the example build is manually invoked. The `${_IMAGE_BASE}` variable contains repository information and is dynamically injecting `${PROJECT_ID}`. `${_IMAGE_URL}`, which is used by the build step, is the combination of `${_IMAGE_BASE}` and `${_IMAGE_NAME}`. This will allow us to substitute repository information, as well as the image name if needed; otherwise, the defined defaults in the configuration will be used.

In the build step, notice that the container image tags are appended, one of which is `${BUILD_ID}`. This is coming from the Cloud Build metadata. There's also `${TAG_VERSION}`, which is a variable.

If this were manually invoked, a straightforward `gcloud builds submit` is all that is needed. If this were triggered, the `dynamic_substitutions` option is not required, and variables can be customized in the trigger if required. After the build is completed, Artifact Registry contains the image and our configured tags:

```
$ gcloud artifacts docker images list us-central1-docker.
pkg.dev/**REDACTED_project_id**/image-repo/nodejs-bookshelf
--include-tags
```

The following is the output from running the preceding command:

```
IMAGE: us-central1-docker.pkg.dev/**REDACTED_project_id**/
image-repo/nodejs-bookshelf
DIGEST:
sha256:aa4ba108423771db58502095fbc8a70a7f54fbb56fdd09b38bbcf78a
7a39bcc5
TAGS: 36e3094e-be72-41ee-8f02-874dc429764e, dev
```

Leveraging variables and dynamic variables can help us create more flexible pipelines and allow teams to adjust values as needed without modifying the build configuration.

Changes in Cloud Build related to secret management

In *Chapter 6, Managing Environment Security*, we covered using secrets while leveraging two different GCP services: **Secret Manager** and Cloud **Key Management Service** (**KMS**). While Secret Manager is the recommended way, Cloud KMS can also be used to declare secrets. The recommended way is to use the `availableSecrets` syntax in your Cloud Build configuration. The `availableSecrets` syntax can be used to declare both Secret Manager and Cloud KMS resources.

The following is the previous declaration method:

```
...
secret:
  - kmsKeyName: projects/**PROJECT_ID_redacted**/locations/
global/keyRings/secret_key_ring/cryptoKeys/secret_key_name
    envMap:
      MY_SECRET: '**ENCRYPTED_SECRET_encoded**'
...
```

The following is the recommended declaration method, as noted in *Chapter 6, Managing Environment Security*:

```
...
availableSecrets:
  inline:
  - kmsKeyName: projects/**PROJECT_ID_redacted**/locations/
global/keyRings/secret_key_ring/cryptoKeys/secret_key_name
    envMap:
      MY_SECRET: '**ENCRYPTED_SECRET_encoded**'
...
```

The `availableSecrets` syntax provides an ongoing way to define secrets and can be used with multiple Google Cloud services.

Configuration parameters within Cloud Build can help organizations prepare for production. While private pools allow for more concurrent builds, it's still important to manage build life cycles. For example, setting expirations when a build has queued for too long is a good habit to have. Variable substitutions provide for more flexible and dynamic build pipelines to help determine the behavior and outputs of builds. The Secret Manager integration with Cloud Build introduced a new syntax that can be used with both Secret Manager and Cloud KMS.

Speeding up your builds

Whether you are just getting started or have been leveraging Cloud Build for a while, it's always good to seek ways to reduce the amount of time it takes for a build to take place. The following is an example list:

- Specifying a `machineType` with more resources, such as `standard`, `highmem`, and `highcpu` (covered in *Chapter 2, Configuring Cloud Build Workers*)
- Parallelizing steps using `waitFor` (covered in *Chapter 4, Build Configuration and Schema*)
- Leaner builder container images for each build step
- Caching intermediate image layers to be used in subsequent builds

The techniques for optimizing Docker images (`https://docs.docker.com/develop/dev-best-practices/`) also apply to Cloud Build builder images. The smaller the images are, the less time it takes for the Cloud Build worker to pull the image and begin executing the step. Having images that separate the dependencies for building application code and the libraries required to execute an application at runtime may be different, so they should be isolated as different images. Removing unnecessary files, package caches, and temp files is also critical to reducing container image sizes.

Caching is another mechanism that can help reduce the time of pipelines that involve Docker container images being built. Cloud Build supports the use of **Kaniko** (https://github.com/ GoogleContainerTools/kaniko) as a tool to build container images instead of the traditional Docker binary. Kaniko allows you to assign a unique key to intermediate steps in a Dockerfile. These keys are used to identify if a container image build step needs to be re-performed because of changes or leverage the cache, which can help improve the performance of a container image build.

In this example, we will use the Node.js bookshelf application (https://github.com/ GoogleCloudPlatform/nodejs-getting-started) in a container image.

The following is the sample Dockerfile (https://cloud.google.com/build/docs/ kaniko-cache#example_using_kaniko_cache_in_a_nodejs_build) that we will use in this example:

```
FROM node:8
WORKDIR /usr/src/app
COPY package*.json ./
RUN npm install
COPY . .
CMD [ "npm", "start" ]
```

Instead of using a typical docker build -t image, we will use a Kaniko executor to specify cache details in this sample cloudbuild.yaml file:

```
steps:
- name: 'gcr.io/kaniko-project/executor:latest'
  args:
  - --destination=us-central1-docker.pkg.dev/${PROJECT_ID}/
image-repo/nodejs-bookshelf
  - --cache=true
  - --cache-ttl=1h
```

The Kaniko executor is run to build a Node.js application. The cached layers are unavailable, and snapshots are taken:

```
...
INFO[0002] No cached layer found for cmd RUN npm install
INFO[0002] Unpacking rootfs as cmd COPY package*.json ./
requires it.
...
INFO[0019] Taking snapshot of files...
```

```
. . .
INFO[0019] RUN npm install
INFO[0019] Taking snapshot of full filesystem...
...
DURATION: 1M
```

In a subsequent run with no changes, the build time is reduced to almost half – from 1 minute to 31 seconds. The cache layer is found and used in the image build:

```
. . .
INFO[0001] Checking for cached layer us-central1-docker.pkg.
dev/**REDACTED_project_id**/image-repo/nodejs-bookshelf/
cache:5bf2e0a45d0dc8dd3109fa3ff73a1ebc195c64b476951c28bf25065
c25250cd4...
INFO[0001] Using caching version of cmd: RUN npm install
INFO[0001] Unpacking rootfs as cmd COPY package*.json ./
requires it.
. . .
INFO[0015] RUN npm install
INFO[0015] Found cached layer, extracting to filesystem
...
DURATION: 31S
```

In this example, Artifact Registry is used to store container images and Laniko creates a nodejs-bookshelf/cache to be used in subsequent builds. We can use the gcloud command to list our images:

```
$ gcloud artifacts docker images list us-central1-docker.pkg.
dev/**REDACTED_project_id**/image-repo --include-tags
```

The following output shows the images in Artifact Registry that contain the built container image, as well as the cache layers created by Kaniko:

```
. . .
IMAGE: us-central1-docker.pkg.dev/**REDACTED_project_id**/
image-repo/nodejs-bookshelf
DIGEST:
sha256:6c31e49a502904fa73a466547876d38d3b6ed39c81a95c9050a2747d
167d1b2f
TAGS: latest
. .
```

```
IMAGE: us-central1-docker.pkg.dev/**REDACTED_project_id**/
image-repo/nodejs-bookshelf/cache
DIGEST:
sha256:687201b973469a84dee8eb5519f9bebf630cd2e8a5599d2579760250
520672b9
TAGS:
5bf2e0a45d0dc8dd3109fa3ff73a1ebc195c64b476951c28bf25065c25250
cd4
```

The image tags for the cached images listed in the preceding output help Kaniko identify if particular container image layers need to be rebuilt.

Many paths can be taken and should be considered to reduce build times overall, which can help improve productivity. Tools should be incorporated to help facilitate overall pipeline time reduction. In this section, we reference Laniko, but code builds can also be improved with **Bazel** (https://bazel.build/). Bazel is an open source build and test tool with a cloud builder image (https://github.com/GoogleCloudPlatform/cloud-builders/tree/master/bazel).

Summary

Cloud Build doesn't differentiate between environments (for example development, non-production, and production), but configurations and recommended practices should be considered to maximize all of its capabilities. In this chapter, we covered topics ranging from optimization, observability, security, and patterns to help with the adoption of Cloud Build within your organization. Each of these topics can be incorporated into your build configurations to satisfy your requirements.

In the next chapter, we will provide a forward-looking view of Cloud Build and look at what is to come.

Part 4:
Looking Forward

In the final part of this book, we will focus on epilogue, covering where Cloud Build as a platform for automation is headed next, its integrations with Cloud Deploy, and running Cloud Build in a hybrid cloud environment leveraging Tekton.

This part comprises the following chapter:

- *Chapter 11, Looking Forward in Cloud Build*

Looking Forward in Cloud Build

Up until now, we have reviewed Cloud Build's foundations, its common use cases, and the patterns for applying Cloud Build in practice.

However, Cloud Build – both at the time of publication and beyond – continues to expand its functionality and integrate more deeply with the rest of the Google Cloud ecosystem.

An example of this includes creating an API for opinionated CI/CD pipelines that uses Cloud Build to execute these pipelines.

In this chapter, we will review some of the areas to look forward to as far as Cloud Build is concerned, specifically covering the following topic:

- Implementing continuous delivery with Cloud Deploy

Implementing continuous delivery with Cloud Deploy

Previously, in *Chapter 8, Securing Software Delivery to GKE with Cloud Build* we reviewed how you could roll out new versions of applications to Google Kubernetes Engine using Cloud Build. When defining these rollouts using Cloud Build, there is very little opinion built into Cloud Build – ultimately, users are responsible for defining each build step, connecting the build steps into a full build, and potentially defining the bespoke build triggers for each build.

One of the key things that users will need to consider is how to promote an application between environments. Environments and their respective functions can include the following examples:

- **Dev**: The development environment in which applications can run, where developers can test their app's dependencies on other services in the cluster that are beyond their ownership. Rather than pulling all other dependencies to their local environment, as with a local Kubernetes cluster, they can deploy to this environment instead:

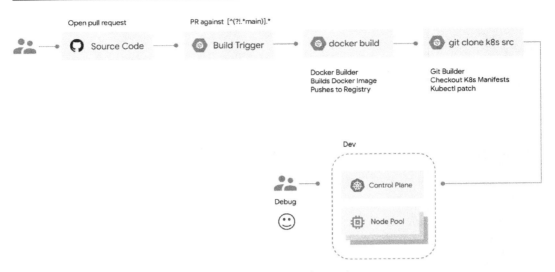

Figure 11.1 – Deploying to a dev environment

- **Staging**: Staging environments are typically close to identical to what will eventually run in production. Ideally, bugs and issues have been identified in earlier stage environments – not only dev but also potentially a dedicated QA or testing environment where more complex test suites such as integration have already passed. In this environment, you may want to perform tests that are closer to the experiences of end users, such as smoke tests. You may also want to implement chaos engineering in environments of this kind, where you can attempt to test the resiliency of your entire system in the event of random outages for various components of your stack:

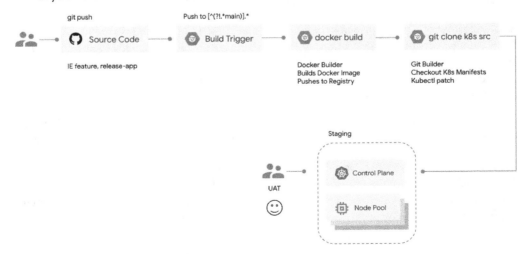

Figure 11.2 – Promoting a release to staging

- **Prod**: Finally, your applications eventually end up in production, where they serve your end users. The rollout to production itself may not be simple, as when deploying to production, you may implement multiple methods for creating the new version of your app and how you shift traffic to that new version:

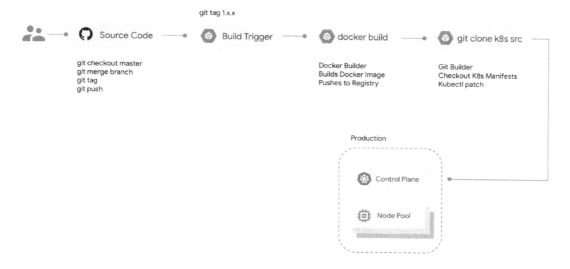

Figure 11.3 – Promoting a release to production

To successfully promote an application to traverse these various environments, it can be challenging to define all of this on your own in Cloud Build. On top of that, if developers define their own builds to accomplish this, there is the chance for drift and silos to occur across your organization.

Therefore, to simplify this increasingly critical process for users, Google has released **Cloud Deploy** – an opinionated model for running templated CI/CD pipelines that is built to roll out an application across multiple target environments that represent the different stages of the software delivery cycle. At the time of publication, Cloud Deploy focuses on serving this use case for users deploying applications to **Google Kubernetes Engine (GKE)** and Anthos clusters.

Cloud Deploy provides a model for using its API to orchestrate the rollout of software to run on Cloud Build workers underneath the hood. Rather than asking users to define each build step along the way, Cloud Deploy exposes resources that allow for declarative pipelines, abstracting away the complex underlying orchestration of Cloud Build resources.

In Cloud Deploy, instead of interfacing with builds and build steps, users instead interface with a combination of Kubernetes and Cloud Deploy specific resources. From a Kubernetes perspective, users will author configurations consistent with the Kubernetes API resources without any required changes.

However, to tell Cloud Deploy which Kubernetes configurations to read, you will author a Skaffold configuration. **Skaffold** is an open source tool that wraps tools such as `kubectl`, manifests rendering tools such as `helm` or `kustomize`, and builds tools such as **buildpacks** to deploy Kubernetes configurations to a cluster.

The following is an example of a `skaffold.yaml` file, which defines the Skaffold profiles to map environments to the respective Kubernetes configuration that will be deployed to them:

```
# For demonstration purposes only, to run this build would
require setup beyond the scope of its inclusion
apiVersion: skaffold/v2beta16
kind: Config
profiles:
  - name: dev
    deploy:
      kubectl:
        manifests:
        - config/dev.yaml
  - name: staging
    deploy:
      kubectl:
        manifests:
        - config/staging.yaml
  - name: dev
    deploy:
      kubectl:
        manifests:
        - config/prod.yaml
```

Users then begin to use the Cloud Deploy-specific resources to define a `DeliveryPipeline` resource. Along with the aforementioned `skaffold.yaml` file, these define the template for delivering a Kubernetes configuration to multiple environments.

A `DeliveryPipeline` resource may look similar to the following configuration in YAML:

```
# For demonstration purposes only, to run this build would
require setup beyond the scope of its inclusion
apiVersion: deploy.cloud.google.com/v1
kind: DeliveryPipeline
metadata:
```

```
  name: app
description: deploys to dev, staging, and prod
serialPipeline:
  stages:
  - targetId: dev
    profiles:
    - dev
  - targetId: staging
    profiles:
    - staging
  - targetId: prod
    profiles:
    - prod
```

Note that DeliveryPipeline outlines the target clusters to which they will deploy the manifests defined in the skaffold.yaml file. It also outlines the order in which the manifests will be applied to each environment, representing the progression of changes through environments such as dev, staging, and prod.

Finally, note that each target also references a profile – this profile aligns to the profiles and specific manifests they align to defined in the skaffold.yaml configuration file. This allows users to perform actions such as flipping the configuration or environment variables between different environments.

Users must then create Target resources referencing each respective GKE cluster and their fully qualified resource name. The following Target resources map to the dev, staging, and prod environments defined in the DeliveryPipeline resource:

```
# For demonstration purposes only, to run this build would
require setup beyond the scope of its inclusion
apiVersion: deploy.cloud.google.com/v1
kind: Target
metadata:
  name: dev
description: development cluster
gke:
  cluster: projects/PROJECT_ID/locations/us-central1/clusters/
dev

---
apiVersion: deploy.cloud.google.com/v1
kind: Target
```

```
metadata:
  name: staging
description: staging cluster
gke:
  cluster: projects/PROJECT_ID/locations/us-central1/clusters/
staging
---
apiVersion: deploy.cloud.google.com/v1
kind: Target
metadata:
  name: prod
description: prod cluster
gke:
  cluster: projects/PROJECT_ID/locations/us-central1/clusters/
prod
```

Now that users have defined their respective targets, they can then begin to use their `DeliveryPipeline` template to create an individual `Release` – which references that `DeliveryPipeline` template and also intakes a `skaffold.yaml` configuration file.

The following command shows how users can kick off by deploying the new version of their application to their desired targets using the `Release` resource:

```
$ gcloud deploy releases create release-01 \
  --project=PROJECT_ID \
  --region=us-central1 \
  --delivery-pipeline=my-demo-app-1 \
  --images=app-01=gcr.io/PROJECT_ID/image:imagetag
```

This resource compiles down into a `Rollout` for each `Target`.

This is where we see a stark difference with doing this via Cloud Build alone. Cloud Deploy enables you to promote an artifact through each `Rollout`, while also providing easy roll-back mechanisms when deploying this artifact to each `Target` environment:

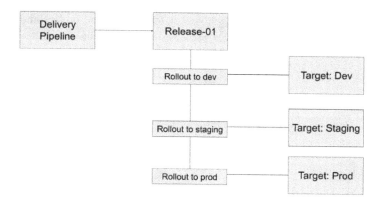

Figure 11.4 – The relationship structure when creating a Release

With the fundamental usage of Cloud Deploy covered, let's now review how Cloud Deploy uses Cloud Build to implement its functionality.

The relationship between Cloud Build and Cloud Deploy

Each `Rollout` from a given `Release` in Cloud Deploy runs Skaffold as a single build step build to deploy the appropriate resources to their respective target environments.

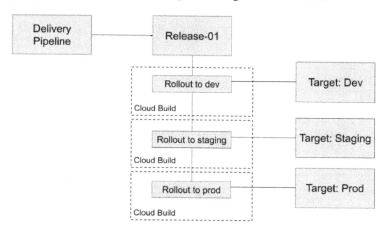

Figure 11.5 – The relationship structure when a Release and Rollouts execute

Users do not need to navigate to Cloud Build – however, to monitor and understand the stage their release is at, rather, the Cloud Deploy console provides pipeline visualization to bubble up this information to users:

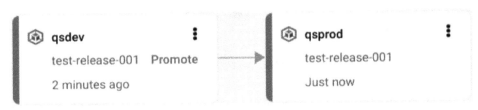

Figure 11.6 – A pipeline visualization in Cloud Deploy

You can use the pipeline visualization to navigate to each rollout and roll back or promote the release to kick off the next rollout.

Because each rollout to a target is its own Cloud Build build, you can also provide the ability to require approvals before rolling out to a target. For environments such as production, where users may want to implement two-person approvals and releases, this is useful.

Kicking off a release can also be set in its own Cloud Build build step, as demonstrated in the following `config`:

```
...
- name: gcr.io/google.com/cloudsdktool/cloud-sdk
  entrypoint: gcloud
  args:
    [
      "deploy", "releases", "create", "release-${SHORT_SHA}",
      "--delivery-pipeline", "PIPELINE",
      "--region", "us-central1",
      "--annotations", "commitId=${SHORT_SHA}",
      "--images", "app-01=gcr.io/PROJECTID/image:${COMMIT_SHA}"
    ]
...
```

This build step can be run in its own build, or it can also be tacked on to a build in which the previous build steps perform the CI and build the container image that will be released via Cloud Deploy.

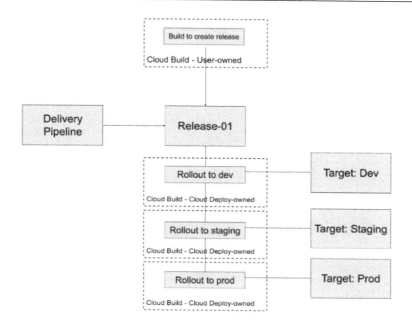

Figure 11.7 – The relationship structure when kicking off a release using Cloud Build

Cloud Deploy, which launched to General Availability in January 2022, is a prime example of an opinionated set of resources that orchestrates the work that Cloud Build can perform. It provides a distinct way to implement patterns while also surfacing a visualization of what can be complex but related tasks.

Cloud Deploy is also reflective of a common pattern – users running multiple GKE environments and clusters. This expands to users running Kubernetes clusters that exist in Google Cloud and elsewhere, such as their own data centers or co-located environments.

Summary

As we look to what's next for Cloud Build, we can appreciate the concept of how abstraction caters to different personas and users. We reviewed how Cloud Deploy provides an abstraction to simplify writing CI/CD pipelines for GKE on top of Cloud Build features such as builds and approvals. Cloud Build will only continue to expand the number of use cases and patterns it supports, enhancing the features we have reviewed in this book, as well as building on the newer focus areas reviewed in this chapter specifically.

Index

`Packt.com`

Subscribe to our online digital library for full access to over 7,000 books and videos, as well as industry leading tools to help you plan your personal development and advance your career. For more information, please visit our website.

Why subscribe?

- Spend less time learning and more time coding with practical eBooks and Videos from over 4,000 industry professionals

- Improve your learning with Skill Plans built especially for you

- Get a free eBook or video every month

- Fully searchable for easy access to vital information

- Copy and paste, print, and bookmark content

Did you know that Packt offers eBook versions of every book published, with PDF and ePub files available? You can upgrade to the eBook version at `packt.com` and as a print book customer, you are entitled to a discount on the eBook copy. Get in touch with us at `customercare@packtpub.com` for more details.

At `www.packt.com`, you can also read a collection of free technical articles, sign up for a range of free newsletters, and receive exclusive discounts and offers on Packt books and eBooks.

Other Books You May Enjoy

If you enjoyed this book, you may be interested in these other books by Packt:

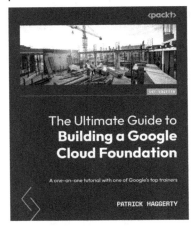

The Ultimate Guide to Building a Google Cloud Foundation

Patrick Haggerty

ISBN: 978-1-80324-085-5

- Create an organizational resource hierarchy in Google Cloud
- Configure user access, permissions, and key Google Cloud Platform (GCP) security groups
- Construct well thought out, scalable, and secure virtual networks
- Stay informed about the latest logging and monitoring best practices
- Leverage Terraform infrastructure as code automation to eliminate toil
- Limit access with IAM policy bindings and organizational policies
- Implement Google's secure foundation blueprint

Google Cloud for DevOps Engineers

Sandeep Madamanchi

ISBN: 978-1-83921-801-9

- Categorize user journeys and explore different ways to measure SLIs
- Explore the four golden signals for monitoring a user-facing system
- Understand psychological safety along with other SRE cultural practices
- Create containers with build triggers and manual invocations
- Delve into Kubernetes workloads and potential deployment strategies
- Secure GKE clusters via private clusters, Binary Authorization, and shielded GKE nodes
- Get to grips with monitoring, Metrics Explorer, uptime checks, and alerting
- Discover how logs are ingested via the Cloud Logging API

Packt is searching for authors like you

If you're interested in becoming an author for Packt, please visit `authors.packtpub.com` and apply today. We have worked with thousands of developers and tech professionals, just like you, to help them share their insight with the global tech community. You can make a general application, apply for a specific hot topic that we are recruiting an author for, or submit your own idea.

Share Your Thoughts

Now you've finished *Cloud Native Automation with Google Cloud Build*, we'd love to hear your thoughts! Scan the QR code below to go straight to the Amazon review page for this book and share your feedback or leave a review on the site that you purchased it from.

`https://packt.link/r/1801816700`

Your review is important to us and the tech community and will help us make sure we're delivering excellent quality content.

www.ingramcontent.com/pod-product-compliance
Lightning Source LLC
Chambersburg PA
CBHW060543060326
40690CB00017B/3589